CLIMATE CHANGE FROM THE STREETS

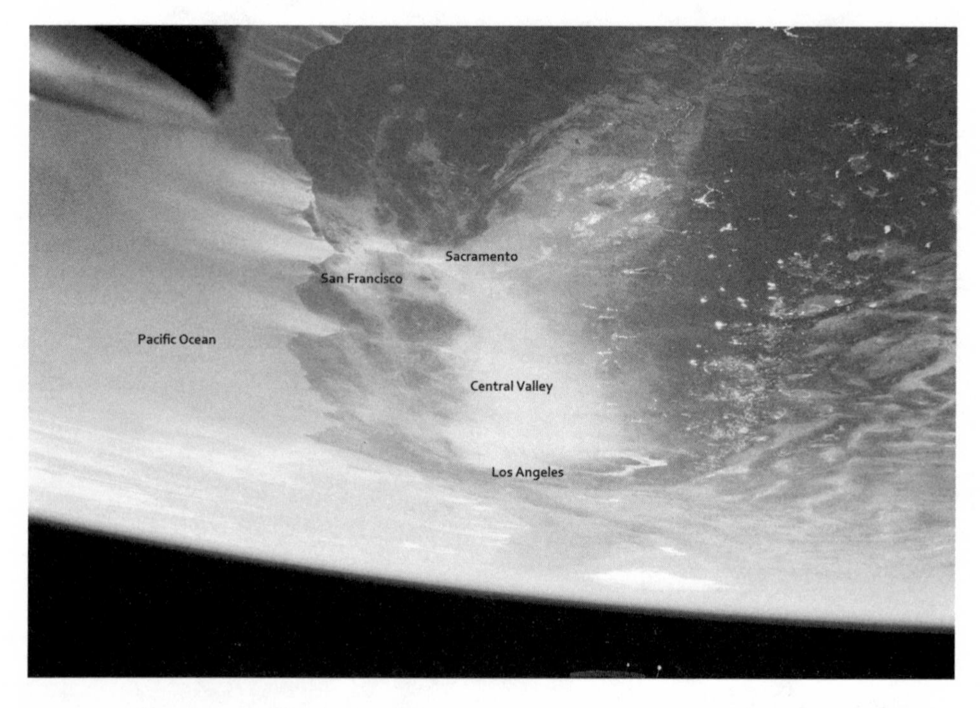

An astronaut aboard the International Space Station photographed the thick layer of smog that hovered over many regions of California in the winter of 2013–14. Courtesy of NASA JSC Earth Science and Remote Sensing Unit

CLIMATE CHANGE FROM THE STREETS

How Conflict and Collaboration Strengthen the Environmental Justice Movement

MICHAEL MÉNDEZ

Yale
UNIVERSITY PRESS
New Haven and London

Published with assistance from the foundation established in memory of Calvin Chapin of the Class of 1788, Yale College, and with the assistance of the Frederick W. Hilles Publication Fund of Yale University.

Yale University Press books may be purchased in quantity for educational, business, or promotional use. For information, please e-mail sales.press@yale.edu (U.S. office) or sales@yaleup.co.uk (U.K. office).

Set in Janson type by Integrated Publishing Solutions.
Printed in the United States of America.

Library of Congress Control Number: 2019940371
ISBN 978-0-300-23215-8 (hardcover : alk. paper)

A catalogue record for this book is available from the British Library.

This paper meets the requirements of ANSI/NISO Z39.48-1992 (Permanence of Paper).

10 9 8 7 6 5 4 3 2 1

Sometimes it is the people no one imagines anything of who do the things that no one can imagine.

—ALAN TURING (attributed)

Contents

Preface

THIS BOOK IS ABOUT people, place, and power in the context of climate change and inequality. Although the science of climate change is clear, policy decisions about how to respond to its effects remain contentious. Even when such decisions claim to be guided by objective knowledge, they are made and implemented through political institutions and relationships—and all the competing interests and power struggles that this implies. I tell a story about the perspectives and influence that low-income people of color bring to their advocacy work on climate change. They are making their voices heard despite the structural disadvantages that often exclude them from public debate. For them, the main threat from climate change is the disproportionate harm that it causes to health in their communities, including an increase in illnesses, injuries, and deaths from extreme events such as hurricanes and wildfires and a rise in respiratory diseases caused by degraded air quality that is further worsened by prolonged drought.

This insight leads them to view climate change as an embodied phenomenon that affects people's daily lives in multiple ways. Recognition of these dynamics is shifting discussions about the broad effects of climate change. Debates are emerging on the unequal harm that it already causes to certain people and communities. This book unpacks the argument that environmental protection and improving human health are inextricably linked, and maintaining that link is key to advancing future climate action policies.

The importance of such links was profoundly apparent to me in the winter of 2013–14, during daily four-mile runs along the Sacramento River, which parallels busy Interstate Highway 5. Over time, those runs began to take a heavy toll on my chest, an effect of the pollution that was accumulating during one of the driest winters on record. From Los Angeles to Sacramento, a haze of gray particles hung in the air. For more than a month, the haze, visible from outer space, obstructed the view of the Sierra Nevada mountain range. Simultaneously, a high-pressure ridge, four miles high and off the West Coast, prevented Pacific storms from cleansing the air. With no rain since early December 2013, pollution levels spiked. In the San Joaquin Valley, California's agricultural heartland, airborne fine particulate matter accumulated to unhealthy levels. Conditions continued to deteriorate until sporadic rain showers came in late January 2014. During these dry winter months, air quality officials declared "red alert days" and warned people to stay indoors. On one red alert day, I ignored the warning and took my daily run. By the third mile, my head had started to ache, and I had begun coughing heavily. Feeling anxious, I stopped to catch my breath and walked the last mile home.

Research by Noah Diffenbaugh and colleagues, published in the *Proceedings of the National Academies of Sciences*, concluded that the state's drought conditions are far more likely to occur in today's warming world than in one without human-caused emissions of greenhouse gases. In light of global climate change, it is projected that by the end of the century California will experience a much greater number and severity of dry, hot-weather seasons. According to climate scientists, the resultant droughts, heat waves, and wildfires can all have a profound effect on air quality. During heat waves and droughts, the air becomes stagnant and traps emitted pollutants, often resulting in increases in surface ozone that can accelerate smog formation. These extreme weather events can also prolong the wildfire season by drying out vegetation and providing more fuel for wildfires, whose smoke is a serious public health threat. State policymakers have referred to these climate change–induced trends as the "new abnormal"—a state of perpetual climate crisis that has the potential to continue for generations.

Red alert days thus represent a rapidly changing climate in California. Air quality warnings were previously issued only in the summer and directed at sensitive populations, such as children, the elderly, and those with respiratory problems. More recently, however, warming temperatures, wildfires, and extreme drought conditions mean increased health risks for everyone. In the San Joaquin Valley, where air quality has deteriorated over recent decades, school officials have long flown colored flags to indicate air conditions: green for good; yellow, moderate; orange, unhealthy for sensitive groups; and red, unhealthy for all population groups. By January 2014, declining air quality forced officials to introduce a new flag color: purple. It indicated "very unhealthy" air for everyone. When a purple alert was declared, schools were forced to fly red flags because they had no purple flags. Until then, such flags had never been needed. The purple alert banned all outdoor activity for teachers and 30,000 students in the Bakersfield City School District. In the district, 91 percent of the population are students of color, 30 percent are classified as English language learners, and almost 90 percent of the students receive free or reduced-price lunches. According to a local elementary school principal quoted in the *Portville Recorder*, that winter was the "first time ever we have been on an inside schedule when it is cold outside. Usually we see this kind of thing when it is hot. Not in January." Yet such poor air quality is not restricted to the San Joaquin Valley. When California entered its third consecutive year of extreme drought in 2014, weather patterns helped create some of the highest levels of soot in the metropolitan regions of Los Angeles and San Francisco.

These smoggy winter skies recall a period in the 1950s and 1960s when Californians endured conditions similar to those now faced by residents of New Delhi or Beijing. Experts predict that extreme weather associated with climate change could reverse decades of efforts to improve the state's air quality. With these implications in mind, one of the oldest and best-known medical journals, the *Lancet*, convened a commission in 2015 to examine the impacts of climate change on human health. That commission found that "climate change is the biggest global health threat of the 21st century." Yet it also concluded that "tackling climate change could be the greatest global health opportunity of the 21st century." This second finding is predicated on the fact that many actions needed to mitigate climate change—reducing fossil fuel use, promoting walkable and bikeable cities, and supporting sustainable transportation and energy systems—offer significant public health benefits. Improvements in the built environment aimed at reducing greenhouse gas emissions can also improve local air quality and lower the incidence of respiratory and cardiovascular illnesses. These illnesses include asthma and heart attacks, which are exacerbated by climate change–related events such as heat waves, droughts, wildfires, and bad air pollution days.

A multibenefit policy approach is important to the contemporary environmental justice movement, which originated from civil rights campaigns organized to expose the socially uneven environmental impacts of industrialization on low-income people of color. The US Environmental Protection Agency has conceptually defined "environmental justice" as the "fair treatment and meaningful involvement of all people regardless of race, color, national origin, or income with respect to the development, implementation, and enforcement of environmental laws, regulations, and policies." Over the last several decades, the movement has expanded in scope, with campaigns centering on a multitude of socio-environmental and land-use issues. Through a uniquely integrated approach, the movement captures the multifaceted interactions of humans and the environment, thus making it relevant across a broad range of issues. Unlike traditional environmental groups, such as the Natural Resources Defense Council and the Environmental Defense Fund, which generally focus on protecting natural resources and wildlife habitats, environmental justice groups center their advocacy on community-specific public health campaigns.

The run I took on that red alert day in Sacramento highlighted for me the ways in which climate change—so often considered a global problem—can have uneven local health impacts. This is a crucial insight that environmental justice activists bring to the public conversation about climate change: pollution most heavily affects the daily experiences and well-being of people living next to noxious sources. The main ingredients of smog (for example, nitrogen oxides, volatile organic compounds, and particulate matter) and the greenhouse gases that cause climate change are often emitted together as fossil fuels burn, and climate change intensifies the localized dangers of polluted air. But climate policies seldom address these interrelated problems jointly or across scales. Environmental justice activists assert that without integrated and contextual policies to address local equity concerns, climate change will likely reinforce and increase current and future health disparities in marginalized communities.

Understanding the vulnerability of communities is important because the risks of wildfires, drought, flooding, and extreme weather associated with climate change bring accompanying impacts to health that are materializing sooner than projected. Location is an important determinant of hazardous exposure, and certain places will see greater harm than others. Research has shown that population groups with low socioeconomic status are more vulnerable to climate change impacts, compared to wealthier groups who can afford the costs of protection against or recovery after such disasters. Seth Shonkoff and his environmental health colleagues refer to such circumstances as the "climate gap," a phrase that aims to capture "the disproportionate and unequal implications that climate change and climate mitigation hold for people of color and the poor."

The threat of global climate change challenges societies to think and act differently to protect communities and the environment. If we are to succeed, we must devote critical attention to the human dimension (for example, local knowledge, culture, and history) of climate policymaking. In California, tensions are particularly evident among various legislators, regulators, business leaders, and activists on these issues. The story of climate change in the state exemplifies the complexity and possibility of this historic moment. California is a world leader in its efforts to address climate change; it is home not only to traditional environmental organizations but also to long-established environmental justice groups that have recently increased their engagement with climate policy. Shifting demographics mean that support from Latina/o lawmakers is now crucial for the passage of environmental legislation. At the same time, industrial and agricultural interests play a strong role in the state's economy and governance. These factors combine to reveal the contested politics of the local, state, and transnational levels on which California makes climate change policy and takes action.

This book project began in my work in politics and environmental policy. My real passion for this research, however, sprang from my experience growing up in low-income, Latino immigrant communities of Los Angeles that faced multiple environmental dangers. As a youth, I was surrounded by people resisting environmental injustice. Whether protesting the siting of landfills or organizing to demand the right to breathe healthy air, they sought to understand how these situations originated, to develop solutions, and to imagine new urban futures. I was inspired by these experiences, and they became an important motivation in my research, teaching, and public service work.

In this book, I take an interdisciplinary research approach that follows the narratives and actions of activists, traditional environmental groups, industry, experts, and governments as they respond to climate change and its effects in California and beyond. My analysis includes nearly fifteen years of firsthand observations (2003–18) that I gathered while working on public policy in the California State Capitol as an advisor, senior consultant, and lobbyist and as a gubernatorial appointee during the passage of the state's internationally acclaimed climate change laws. In those roles I gained valuable insight into ways in which governments, businesses, and nongovernmental organizations interact to shape climate policy. My connections with environmental justice groups in particular helped me understand how community and public health benefits can be more effectively integrated with global climate change policy. Although most scholarship in the field of environmental studies has focused on elite actors (such as policymakers, traditional environmentalists, scientists, and industry), my research provides a more nuanced analysis of the influence of environmental justice activists in transforming climate change policies. In this book, I seek to amplify their efforts and voices, which have largely been ignored in the narrative of California's global leadership on climate change—the voices and stories of people such as Mari Rose Taruc, Sofia Parino, Emily Kirsch, Martha Dina Arguello, Brian Beveridge, and Katie Valenzuela.

In *Climate Change from the Streets*, I analyze how environmental justice advocates have strategically engaged in the policymaking process to orient climate policies toward public health outcomes at multiple scales (see the timeline and map in the Appendix). My account weaves together analyses of three interconnected case studies: (1) climate and public health activism in two heavily impacted communities of color (Richmond and Oakland); (2) conflict over state-level carbon trading and use of its revenue for investment in the local communities most harmed by air pollution; and (3) international and local implications of the forest conservation projects in the Global South (Mexico and Brazil) allowed under California's market-based climate laws. These cases shed new light on the links between climate change and local environmental inequities, as well as the effects of climate policy and action at multiple scales. They highlight the diverse forms of knowledge that environmental justice activists use to challenge conventional policy solutions. These activists are reconceptualizing the meaning, scope, and scales of climate change to redress underlying environmental inequities in communities globally.

Through these case studies, I examine the dilemmas that policymakers and activists face as they seek to address these problems, sometimes in collaboration but often in conflict with each other. This book therefore makes three major contributions to our understanding of inequality in the context of climate change. First, I demonstrate that public health and environmental justice perspectives can be central to successful climate policy development and implementation. Second, I offer an interdisciplinary framework for theorizing the kinds of negotiations between scales and worldviews that are involved in the development of equitable climate change policy. Finally, I provide a set of findings that activists can use to negotiate with governments that legitimizes their perspectives about the differential impact of climate change on disadvantaged communities.

Activists living next to polluting sources have moved from the margins to the center of global environmental policies. They represent groups rooted in some of the nation's poorest neighborhoods, most directly affected by climate change and pollution. Through advocacy campaigns, community-based research practices, and lawsuits, these activists have transformed environmental protection paradigms by insisting upon the importance of their own "embodied perspectives." Throughout this book, I document how individuals and activist groups have organized to ensure that climate change solutions tackle both global problems and local needs. I offer their example as a critically important case study for scholars, policymakers, advocates, urban planners, and environmental analysts seeking new directions in climate policy and justice worldwide.

Acknowledgments

Climate Change from the Streets would not have been possible without the efforts of the activists and policymakers who generously shared their struggles and stories with me. These include Mari Rose Taruc, Catalina Garzón, Emily Kirsch, Brian Beveridge, Nidia Bautista, Sofia Parino, Susana De Anda, Laurel Firestone, Jeff Conant, Martha Guzman, Jonathan Nelson, Dean Florez, Phil Serna, Mary Nichols, Ricardo Lara, Cristina Garcia, Richard Corey, Bill Magavern, Fabian Núñez, Kevin de León, Gladys Limón, Eduardo Garcia, and Cindy Montañez. I followed them and other actors who have asked to remain unnamed through intersectional approaches and multiple scales. Through them I understood the diverse ways in which environmental justice groups—with roots in low-income communities of color and Indigenous and immigrant communities—have challenged the global conceptualization of climate change to include embodied knowledge and lived experience. Activists, moreover, have shown me how they operationalize an embodied approach to develop alternative climate change futures. I am continually inspired by their work and dedication to seeking a more just world.

I am also indebted to a supportive intellectual community throughout the country: in particular, scholars at the University of California system, including Julie Sze, Alastair Iles, Jonathan London, Jason Corburn, Deborah McKoy, and Ben Crow; and at the University of San Francisco, Alice Kaswan, Steve Zavestoski, Jack Lendvay, Mary Wardell-Ghirarduzzi, and the Gerardo Marín Postdoctoral Fellowship program. Special thanks to Sir Peter Crane and

Indy Burke at Yale University, who gave me an amazing opportunity to complete this book project through the Presidential Visiting and Pinchot Fellowship programs. I would like to thank my colleagues at the Yale Center for the Study of Race, Indigeneity, and Transnational Migration: Steve Pitti, Albert Laguna, Laura Barraclough, and Alicia Schmidt Camacho. They provided me with constant encouragement to unpack the rich ethnographic stories and motivations of environmental justice activists. They gave me inspiration to keep moving forward and courage to tell the story that needed to be told.

I am grateful to David R. Diaz, Malo Hutson, Martha Groom, David Winickoff, Carolyn Finney, Daniel Esty, and the entire Ford Foundation Fellowship program for their mentorship. I acknowledge Sheila Jasanoff, David Pellow, Dorceta Taylor, Julian Agyeman, Laura Pulido, Robert Bullard, Jill Harrison, Isabelle Anguelovski, Gwen Ottinger, Harriet Bulkeley, Rachel Morello-Frosch, and Manuel Pastor for their significant body of research, which gave me a strong foundation from which to develop new conceptual models.

This book would not have seen the light of day without the early, continuous, and enthusiastic support of Jean Thomson Black at Yale University Press. She gave me the confidence to understand the bigger picture of this project. I am also thankful to Michael Deneen at the press, for his insightful editorial assistance. Darnell Grisby, Florence Grant, and James Benton were my emotional and editorial sounding boards. I could constantly depend on their foresight, wit, and craftiness to push this book to new levels. Thank you to Timmons Roberts and Damian White for hosting a transformative book workshop at Brown University and to the *Journal of Planning Education and Research* for workshopping portions of the book at the Writing Workshop for Early Career Planning Scholars. Some of the ideas in Chapter 4 were originally published in the book *Spatializing Politics*. I am also grateful to the Yale Whitney Humanities Center and the committee members of the Frederick W. Hilles Publication Fund for their support in the publication of my book. I acknowledge Bill Nelson for his assistance in reproducing many of the book's graphics.

I owe thanks to David Pellow, Debra Kahn, David Diaz, Mari Rose Taruc, Charles Lee, Monamie Bhadra, Manuel Pastor, Jade Sasser, Clint Carroll, Michael Hooper, Paul Sabin, Courtney Knapp,

Irene Garza, Daniel Esty, Timmons Roberts, Damian White, Katie Valenzuela, Alice Kaswan, Steve Zavestoski, Justin Farrell, Bill Funderburk, and Tomo Sasaki, who offered editorial comments that helped expand my scope of knowledge to develop more nuanced stories of California's changing climate. I am also grateful to the anonymous reviewers for their thoughtful and constructive suggestions. I would like to acknowledge the students in my 2018 and 2019 Yale University graduate seminar, Environmental Justice, Nature, and Reflective Practice, who gave me the emotional power to finish this book project. They all motivated me to reflect on my own scholarship and environmental practice and to remember why I am passionate about this field of study.

I am also thankful to my sister Maggie Méndez's fourth-grade class at Bellingham Elementary School in North Hollywood (a predominately Latino immigrant community) for selecting the photo for my book cover. The students chose the photo during their spring 2019 module on climate change. They are part of a new generation of students and future leaders who are gaining environmental literacy through an integrated core subjects curriculum. This is critically important in places such as the Los Angeles Unified School District, where students and families embody a range of cultures and heritages. Over 92 languages are spoken in district schools, which represents the rich diversity of California and the United States.

Finally, I am grateful to my family, friends, and God for providing me with the strength, faith, and inspiration to complete this research. Over the course of this journey, I have faced career changes, confronted doubts, and overcome vast challenges. I could not have completed this endeavor, in particular, without the support and encouragement of my loving parents, Ana and Rodolfo Méndez. Whether near or far, they always provided an intellectual and supportive environment in which to carry out my studies. They gave me my first job, working in their Rudy's Bike Shop in the low-income community of Pacoima—the first Latino-owned bike shop in the San Fernando Valley. For more than 25 years, they struggled to keep the shop afloat to support sustainable and affordable transportation options for low-income immigrant families in the region. This book is dedicated to their efforts as immigrants, green entrepreneurs, and advocates for a more just and sustainable society.

Abbreviations

AB 32	Assembly Bill 32: Global Warming Solutions Act of 2006
Cal. Chamber	California Chamber of Commerce
CARB	California Air Resources Board
CBE	Communities for a Better Environment
CEJA	California Environmental Justice Alliance
CMTA	California Manufacturers and Technology Association
CO_2	Carbon dioxide
EAAC	Economic and Allocation Advisory Committee
EDF	Environmental Defense Fund
EJ	Environmental justice
EJAC	Environmental Justice Advisory Committee
EPA	Environmental Protection Agency
GHG	Greenhouse gas
IPCC	Intergovernmental Panel on Climate Change
NGO	Nongovernmental organization
NO_2	Nitrogen dioxide
NRDC	Natural Resources Defense Council
O_3	Ozone
OCAC	Oakland Climate Action Coalition
ODPHP	Office of Disease Prevention and Health Promotion
OECD	Organisation for Economic Co-operation and Development

OEHHA	Office of Environmental Health Hazard Assessment
PM	Particulate matter
REDD	Reducing Emissions from Deforestation and Degradation
SO_2	Sulfur dioxide
UN	United Nations

CHAPTER ONE

Seeing Carbon Reductionism and Climate Change from the Streets

Climate change is commonly envisioned as a global issue, present in no particular place because it exists everywhere. Experts discuss its workings in abstract, technical language and often picture its effects in terms of melting ice caps, incremental rises in sea level, and damage to ecosystems far from the cities and towns where an increasing proportion of human beings live. Yet these concerns can have little day-to-day relevance for low-income communities of color, which face the severest consequences of a changing climate. This disconnect presents a challenge for environmental justice advocates, whose campaigns are often focused on pollution at the neighborhood level. They have long known how to effectively fight for change and make their voices heard in local government and policymaking. But climate policies at the global scale are crafted in different settings, follow different forms of action, and work toward different goals. In order to participate in decisions about climate change that will inevitably affect marginalized communities, environmental justice activists must adapt their perspectives and strategies from the local scale to the global. How to take interconnected action across scales has become a central concern for them.[1]

To follow environmental justice activists' efforts as they move among these various settings, I use a conventional definition of "scale" as "the spatial reach of actions" and analyze climate change through an approach known as multiscalar analysis.[2] This approach brings into focus the multiple scales (that is, locally, regionally, statewide, nationally, and internationally) in policymaking and environmental justice advocacy as a way to understand how climate change is defined as a problem and how societies seek to act upon it. Seeing a problem at any given scale involves decisions—conscious or not—about which of its aspects to disregard and which to act upon. The key point is that scale is not objective. It is constructed through human relationships and is an important factor in political strategies and contests over power and authority. Changing the scale at which a problem is addressed can alter the power relationships that surround it—the associations that determine unequal access to resources and institutions and the ability to choose and act despite resistance from others.[3]

Any account of climate policy that focuses on a single scale can thus only be partial. Analysis at the global scale will inevitably gloss over the question of how public definitions of climate change can reinforce existing local inequalities in power, resources, and health. If climate change is seen solely as a global phenomenon, then it will seem self-evident that only "global actors"—national governments, the United Nations, multinational corporations, the international community of scientists—are empowered to address it. By seeking to understand the effects of climate change and climate policy at multiple scales, we can promote more egalitarian forms of public decision-making about this critical problem. Coalition-building between distant locations is central to this project. Activists' movement between scales is not just movement between locations; they expand the reach of their actions by developing relationships around common concerns.[4]

Instead of attempting to impose a single conceptual model on these activists' efforts, I draw on a range of ideas and research to unfold how they work in practice. This book contributes to an emerging body of interdisciplinary social science scholarship addressing environmental justice across multiple scales. It also draws on important work in the fields of human geography, critical policy studies, anthropology, science and technology studies, public health, and urban planning, all of which recognize the importance of multiscalar approaches. For example, David Pellow, who has called for a movement toward critical environmental justice studies, emphasizes the political use of scale. He argues that traditional environmental justice literature tends to examine cases at one scale or another, rather than exploring "how EJ struggles function at multiple scales, from the cellular and bodily level to the global level and back." According to Pellow, scale is of critical importance because it shows how activists' responses to environmental injustices draw on spatial thinking, social networks, and knowledge to connect hazards in one place to harm in another. Multiscalar approaches in research "allow us to comprehend the complex spatial and temporal causes, consequences, and possible resolutions of environmental justice struggles."[5]

Scholars in these disciplines, moreover, have called into question the notion of the "field site" in social science research—the location in which the social processes under investigation occur. These researchers are rethinking this basic concept in their disciplines in order to tackle emerging social issues that extend beyond a single site with defined boundaries. An increasingly globalized world necessitates focus on the ways in which people, objects, and ideas are interconnected and move across space and time. Movement is central to social practice, and coherent cultural processes (including policymaking) can occur across great distances, linking up distinct actors. A multiscalar approach, therefore, traces how knowledge travels and is interpreted and reinterpreted along the way.[6] The human geographer Mike Hulme argues against reifying what climate change means in the physical, absolute sense and instead seeks to understand its meanings in different places, to different peoples, and at different times.[7] In the same spirit, through a series of case studies, I follow people, their controversies, and their relationships across space and time, through settings that are related yet spatially noncontiguous.[8] From the streets of Oakland and Richmond to the halls of the California State Capitol and the United Nations to the tropical forests of Mexico and Brazil, environmental justice groups travel between scales to contest or legitimize California's globally linked climate policies. Such groups exemplify the diversity of knowledges and practices around climate change as they engage in policymaking within California and beyond its borders.

Why California?

Cities and states, rather than national governments, are increasingly important as sites for experimentation in climate change mitigation and adaptation. They also represent key sites for environmental justice debates.[9] As the world's fifth-largest economy and the only US state to implement a comprehensive program of regulatory and market-based mechanisms to reduce greenhouse gas emissions, California has consistently been at the forefront of broader national and global environmental experimentation.[10] California's successes in climate change policy at the broadest scales make it an ideal place to investigate the local effects of such decision-making and the ways in which environmental justice concerns have been adopted or excluded in the process.

The passage of Assembly Bill (AB) 32, the Global Warming Solutions Act of 2006, made the state an international leader on climate change science and policy innovation. The state's cap-and-trade program—a central, market-based mechanism for ensuring emissions reductions—is the third largest in the world, after those of the European Union and China. The program establishes a statewide emissions limit—the cap—which is imposed by mandating regulated businesses to either reduce their emissions or purchase permits in order to pollute. There are only as many permits as there is capacity under the cap. Businesses can also trade the permits, generating a market price on greenhouse gas emissions. Supporters of cap-and-trade argue that it reduces emissions at a much lower cost than other mechanisms. At the same time, environmental justice advocates focus attention on underlying equity issues and question the effects of cap-and-trade across different locations and demographic groups, including people of color and low-income communities.[11]

Such questions are important to activists because while California has led the nation in climate change policy, it has also been challenged by inequality. Despite its robust high-tech economy, California also has one of the nation's highest poverty rates. Between 1980 and 2010, the incomes of California families in the poorest 10th percentile dropped 24 percent; meanwhile, the incomes of the wealthiest 10th percentile grew 34 percent. If race and gender are factored in, such trends become even more pronounced. Decades of studies show that communities of color and low-income neighborhoods are disproportionately affected by air pollution from fossil fuel combustion relative to the overall population.[12] A recent study conducted by scientists at the US Environmental Protection Agency and published in the *American Journal of Public Health* found that areas most affected by particulate matter emissions showed significant disparities between communities differentiated by race and income. According to the study, African Americans shouldered the greatest impact, with a 54 percent higher health burden compared to the overall population. Communities of color in general had a 28 percent higher health burden, and those living below the poverty line had a 35 percent higher burden.[13]

Environmental justice advocates have stressed that inequalities within California make linking local equity concerns with state-level climate policy design especially important. They argue that there is additional pressure on policymakers to ensure that climate change solutions provide multiple benefits and prioritize communities living next to large polluters, such as oil refineries and power plants. For environmental justice advocates, the fight against climate change is intertwined with the provision of opportunities for equal access to clean air and jobs. What was once a movement responding to site-specific hazards has started to engage in a much broader and proactive approach at the state level. In addressing climate policy, the environmental justice movement is seeking changes on issues that have typically been the exclusive domain of traditional environmental organizations. Unlike comparable groups in other states, California's environmental justice movement is becoming a sophisticated and key political constituency in the state's environmental politics and is active in shaping legislation and regulation. This perspective is confirmed by a 2017 California Chamber of Commerce report that outlines the growing influence of the environmental justice movement: "The concerns raised by environmental justice groups are ones that the business community must consider. . . . [I]n recent years they have become more organized and thus have a large advocacy presence both in the legislature and regulatory agencies."[14]

The relationship between climate action and environmental justice in California was further underscored at the 2015 United Nations Climate Change Conference in Paris, France. At the invitation of Christiana Figueres, the UN climate chief, California governor Jerry Brown and a delegation of state and local policymakers traveled to the conference to further deepen the state's global partnerships. They received a major platform to articulate how subnational (state, local, and regional) governments could implement local climate programs.[15] The conference resulted in the passage of the Paris Agreement, which committed more than 190 countries to actions and investments for a low-carbon future. It was particularly powerful in acknowledging the role and responsibilities of subnational governments in the fight against climate change, as well as the unjust burdens of climate change borne by the world's most disadvantaged communities. Governors and mayors assembled alongside the national negotiators and declared their readiness to tackle climate change. Their involvement is important because cities account for more than 70 percent of energy-related carbon emissions—a proportion that is expected to grow as urbanization continues worldwide.[16]

Since 2006, California has formed alliances with states and provinces worldwide in an effort to bypass federal inaction and forge direct agreements with foreign partners more motivated to fight climate change.[17] Governments around the world are looking to California for lessons on developing cost-effective greenhouse gas reduction policies that also benefit and safeguard low-income communities of color. The 2016 election of Donald Trump as US president, with his refusal to uphold the Paris Agreement or even federal environmental laws, places a far greater emphasis on state- and city-level action.[18] Within a context of increased anti-environmentalism, understanding the ways in which environmental justice and local climate action are defined, researched, and implemented is especially important.

Whereas no government or community group alone can halt climate change, California's environmental policies have a long history of success and replication. Many of these policies are the result of community residents at the frontlines of pollution resisting, protesting, and collaborating with the state to force the development of more equitable environmental policies. The state's leadership on climate change thus highlights the ability of subnational governments to link local concerns with global forums. It is undeniable that the state's climate change programs are fostered by certain conditions of privilege—a robust economy, racial and ethnic plurality, and progressive statewide leaders. Nonetheless, they offer clear models of how advocacy and policy can cross multiple scales and address the ways in which climate change intersects with social categories such as gender, race, class, Indigeneity, immigration status, and other aspects of human identity.

In the social sciences, the concept of intersectionality has been used to highlight how these categories of culture and identity overlap, which heightens the effects of discrimination, exclusion, social inequality, and systemic injustice in the lives of specific individuals. An intersectional approach to climate change emphasizes how certain people and groups suffer worse effects because of overlapping factors that are often measured separately.[19] California's experience points to ways in which deliberately crafted climate policy can alleviate, rather than reinforce, these harmful patterns.[20] It shows how definitions of effective and just climate action can develop from negotiations among policymakers, experts, and diverse communities. Moreover, California provides other states with a framework for developing political support by linking solutions with larger socioeconomic and environmental issues and focusing on the concrete, local benefits that this integrated type of climate change policy can bring—cleaner air, green-collar jobs, and healthier neighborhoods. Simply put, California tells a story of what's possible if we imagine politics differently.

Until now, there has been no book-length analysis of California's environmental justice activism in the context of climate change. A central objective of *Climate Change from the Streets* is to understand the ways in which the phenomenon has challenged the environmental justice movement's organizing tactics and campaigns. This research builds on Jason Corburn's *Street Science*. Like Coburn, I examine how policymakers and regulators are being pressured through various community-level organizing strategies to discover new ways of combining the expertise of professionals with the embodied knowledge of environmental justice activists. In this sense, I show how climate change policy can shift away from a solely "expert-driven" process and toward a more "street-level" perspective.[21] Yet this study focuses not only on community activism but also on how it can benefit from multiscalar and intersectional approaches. I demonstrate how connecting climate action to key issues such as air quality, women's reproductive rights, immigration, access to employment, and affordable housing can build broad and powerful coalitions. I examine through a series of case studies how climate policy is configured throughout California and is influenced by diverse ways of understanding. These case studies identify four key elements of the relationship between climate change and environmental justice: broad participation in decision-making; an experimental approach to consensus-building; shifting the central goal of climate policies; and the importance of scale in policymaking.

Experiments and Grassroots Climate Action

For more than a decade, California has provided a test case that shows other governments how disparate interest groups—environmental justice advocates, lobbyists, industry, labor unions, policymakers, and scientists—can collaborate on comprehensive climate action policies. Many of these negotiations can be contentious. The interests of participating groups and individuals are often in direct competition, and the path to agreement unclear at the start. Participants' ideas of what climate change solutions should entail are shaped by diverse, and often unspoken, worldviews. Nonetheless, they have the potential to reshape climate futures and power relations in society.[22]

Climate policymaking thus can be seen as a contingent process in which problems and solutions take shape together. To succeed, climate policy needs broad-based support that stems from social and cultural values, as well as trust in science. Yet scientific knowledge alone cannot resolve the problem, and the Intergovernmental Panel on Climate Change's 2014 assessment report acknowledges the importance of local perspectives: "Vulnerability to climate change, greenhouse gas emissions, and the capacity for adaptation and mitigation are strongly influenced by livelihoods, lifestyles, behavior and culture. Also, the social acceptability and/or effectiveness of climate policies are influenced by the extent to which they incentivize or depend on regionally appropriate changes in lifestyles or behaviors."[23]

Grassroots coalitions are pushing local governments to connect climate policy with public health and community benefits. This pressure is creating experimental spaces where elite global actors are working with local activists to broaden perspectives on climate change. These community-based collectives are inspiring local, state, and translocal coalitions to imagine various relationships among the atmosphere, economic inequalities, racial disparities, and climate policy. According to the sociologist Theda Skocpol, such collaborations are important. In her analysis of the failed 2010 push for cap-and-trade legislation in the US Congress, she concluded that traditional environmental organizations, such as the Natural Resources Defense Council, were overly focused on making an insider deal with business interests, with little grassroots support: "To build leverage with Congress, and to push back effectively against elite and populist anti-environmental forces, global warming reformers must mobilize broad, popularly rooted support for carbon-capping measures that have something concrete to offer not just to big corporate players, but also to ordinary American citizens and to local and state groups."[24] It is increasingly clear that community-based forms of action have enormous political potential and are key to the passage and successful implementation of climate policies.

Environmental governance therefore entails important decisions about who makes particular policy choices and which criteria are used in making those choices. Investigation of how climate policy is developed in practice reveals the particular agendas, politics, motivations, and expertise that shape solutions and environmental justice outcomes. It also reminds us that social movements are more than a single nongovernmental organization or group of activists. They are complex networks of members acting together to generate considerable research outputs and political and media pressure to enact change.[25]

Community-based solutions are often characterized as fragmented or lacking rigor. But I argue in this book that they are a significant part of the public response to climate change. They serve as experiments from the streets, where "experiment" means the "action of trying anything, putting it to proof," as well as "to have experience of . . . to feel." Such experiments are key to collective decision-making about climate change and about who has the authority to test new solutions. As urban planning scholars have reasoned, seeing community-based solutions as experiments reveals how and why climate policy travels across scales and becomes institutionalized.[26] Through these experiments, we can assess the interconnection of climate change and politics, tracing how, in public life, scientific facts are never disconnected from social commitments and worldviews. A focus on experimentation provides a critical lens to grasp the ways in which science can stabilize the existing social order and alternative knowledge practices can promote social change.[27]

The US Environmental Justice Movement in the Context of Climate Change

Throughout this book, I provide an analysis of conflict and collaboration in California's policymaking process. I show how groups such as the California Environmental Justice Alliance, the Asian Pacific Environmental Network, the Center on Race, Poverty, and the Environment, and the Oakland Climate Action Coalition are pushing the state and its cities to the forefront on climate policy and experimentation. The work by California's activists provides lessons on both the promise and the limitations of integrating environmental justice into climate policy. I analyze recent climate change advocacy as an evolution of long-standing political and conceptual frameworks. I use the term "environmental justice" instead of "climate justice" to honor this history of community-based forms of action.[28]

The environmental justice movement emerged nationally in 1982, through a protest against the siting of a hazardous waste landfill in Warren County, North Carolina, a rural and predominantly African American region. Although the Warren County protest failed to prevent the siting of the landfill, it was the first environmental protest by people of color in the United States to garner national attention. In California, similar campaigns have opposed industrial contamination; fought mining on Indigenous land; sought meaningful participation in land-use decision-making and chemicals policy reform; and called for stronger regulation of pollution from ports, railroads, and commercial infrastructure.[29] One pioneering effort, the United Farm Workers' antipesticide campaign in the San Joaquin Valley from 1965 to 1971, linked protests over farm labor conditions by publicizing the health effects of pesticides. In 1988, Kettleman City, already host to one of the nation's largest hazardous waste landfills, became the proposed site for a toxic waste incinerator. The predominately Latina/o residents, who were unable to substantively engage in public hearings and deliberations due to the structural racism embedded in those processes, organized a three-year campaign that permanently halted the siting of the incinerator.[30]

The environmental justice movement's earliest claims can be summed up with the concept of "environmental racism," which emphasizes how low-income and racial minority communities are intentionally or unintentionally targeted for disproportionate exposure to pollutants or degraded environments, compared with the general population. This process is argued to be coupled with the systematic exclusion of minorities in decisions on environmental policymaking, enforcement, and remediation.[31] Studies in the 1980s supported this concept by identifying race as a factor in the siting of hazardous waste facilities and toxics-producing facilities in primarily poor communities of color. In 1983, at the behest of members of the Congressional Black Caucus, the US General Accounting Office conducted the first federal study on race and pollution. It found that in the southeastern United States, three out of four commercial hazardous waste landfills were sited in predominately African American communities.[32]

Expanding on this study, in 1987, the United Church of Christ's Commission for Racial Justice conducted a national analysis. It found that "three out of every five Black and Hispanic Americans lived in communities with uncontrolled toxic waste sites."[33] Later research by the sociologist Robert Bullard detailed discriminatory solid waste siting practices in African American communities throughout the southern parts of the United States. These three studies were influential in forging an explicit link between environmental research and social movements.[34]

In 1991, the foundational document of the movement, *Principles of Environmental Justice*, was composed at the First National People of Color Environmental Leadership Summit, held in Washington, DC. It asserted that all people have a right to clean air, water, land,

and food and called for environmental policy based on mutual respect, free from discrimination or bias. Delegates declared communities' right to self-determination and to participate as partners in every level of decision-making, from assessment and planning to implementation, enforcement, and evaluation. Central to these principles was the expanded concept of "environment" beyond ecological and natural systems to include places where people live, work, play, worship, and go to school.[35] These declarations represented the common concerns of diverse social groups attending the summit, including civil rights, farmworker justice, and Indigenous rights organizations.[36] They underscored issues of unequal recognition and participation, in addition to the distribution of environmental harms, and reflected the broader civil rights principle that "race-conscious policies and practices are necessary, specifically to target and address the sources and causes of racial disparities."[37]

Scholars including David Schlosberg have suggested that the concept of environmental justice has four interconnected components. The first, distributive justice, is the notion that environmental burdens and benefits should be allocated equitably. The second, procedural justice, refers to the rights of all people to participate meaningfully in decisions that affect their lives. The third, recognition justice, moves the groups most affected by pollution from the margins toward the center of public discourse. It highlights the lack of recognition of cultural identity and difference and the exclusion from the political process this engenders. The fourth argues that justice also requires the building of capabilities—that is, the basic resources and opportunities that people need to be full members of society. These include access to living-wage jobs, affordable housing, transportation, healthy foods, and healthcare. Over the years, the broadest goal of the environmental justice movement has been to integrate these four components of justice into environmental solutions.[38]

The First National People of Color Environmental Leadership Summit, moreover, catalyzed a series of events at the federal level that culminated in President Bill Clinton's issuance of Executive Order 12898, "Federal Actions to Address Environmental Justice in Minority Populations and Low-Income Populations," in 1994. The order charged all federal agencies with integrating environmental justice concerns into their operations. Following this example, in 1999, California became the first state in the nation to codify a definition of environmental justice in law.[39] The sociologist Manuel Pastor argues that California's environmental justice laws are politically rooted, with Latina/o politicians in particular responding to their core constituencies: "Environmental inequity has been a key concern of Latina/o lawmakers who are new to power but old to pollution, and happen to represent critical swing voters in state elections."[40]

Schlosberg and Lisette Collins recognize the founding of the Environmental Justice and Climate Change Initiative in 2001 as a historic moment linking environmental justice and climate change policy. It resulted from the first Climate Justice Summit at The Hague during the UN climate negotiations the previous year, when a confluence of events and reports highlighted the inequitable impacts of climate change on low-income groups and communities of color globally.[41] Hurricane Katrina in 2005 solidified the relevance of environmental justice concerns in the context of climate change by exposing preexisting injustices in the city of New Orleans. These included racial segregation, poverty, failing schools, and poor housing. It highlighted how communities of color were disproportionately unprepared and vulnerable to the storm. It also demonstrated how these communities received less government support for recovery and were subject to continued forms of discrimination in rebuilding efforts. The disasters of Katrina and extreme weather events in subsequent years became a rallying call as the environmental justice movement in the United States turned its attention more fully to the dangers of climate change.[42]

Civic Epistemologies of Climate Change

Environmental justice activists often influence policy by learning how to "become" environmental health practitioners and city planners when they interact with highly specialized knowledge and practices.[43] These new forms of knowledge and governance can be understood as an articulation of *civic epistemology*, "the institutionalized practices by which members of a given society test and deploy knowledge claims used as a basis for making collective choices."[44]

Originating in the work of the science and society scholar Sheila Jasanoff, the concept of civic epistemology was developed to analyze the practices, methods, and institutions by which a society identifies new policy issues, generates knowledge relevant to their resolution, and puts that knowledge to use in policymaking. Through the cases of biotechnology and climate change, Jasanoff explains the different ways in which citizens in Germany, Britain, and the United States come to know things in common and apply their knowledge to the conduct of politics.[45] Other scholars working with the concept have found that national characteristics such as administrative and legal codes and styles—as well as culturally specific conceptions of risk, vulnerability, and impact—shape scientific definitions of and policy responses to climate change. Thus, scientific studies deemed reliable and legitimate in one country may be dismissed as inadequate for policy guidance in another, even when similar political and economic variables influence regulators in both countries.[46]

Focusing on *multiscalar* civic epistemologies, I argue, can elucidate the interconnected scales and decision-making processes of climate policy, which are different from those that occur solely at the national level. It also assists in understanding how the construction of climate change travels between scales and is influenced by local knowledge, culture, and history. Embeddedness and rootedness are not explicitly acknowledged in Jasanoff's concept of civic epistemology. Her national approach privileges knowledge production by elites without acknowledging that local and community-based activists can influence decision-making on multiple scales. Nor does the national approach analyze how race, gender, class, and power differentiate and shape civic epistemologies.

A multiscalar approach to civic epistemology can reveal whether the distribution and effectiveness of policy are equitable across interconnected scales—global, regional, national, and local. Single-scale analysis obscures the fact that numerous actors travel across scales in relation to environmental justice and climate change.[47] Scientific knowledge and climate policy may remain relatively similar across expert communities globally. But when scientific knowledge is used at different scales and in diverse geographic contexts, the democratic procedures of regulatory practices, public participation, and legitimating science differ greatly.[48] By examining the ways of knowing that inform civic action, we see that diverse publics—not just scientists, urban planners, and policymakers—define, measure, and govern environmental health and climate change.

Climate Change Worldviews

Whereas civic epistemologies are the culturally specific ways in which publics expect the government's expertise, knowledge, and reasoning to be produced, tested, and put to use in decision-making, they are also highly dependent on stakeholders' worldviews.[49] According to the political scientist Michael Lind, "a worldview is a more or less coherent understanding of the nature of reality, which permits its holders to interpret new information in light of their preconceptions. Clashes among worldviews cannot be ended by a simple appeal to facts. Even if rival sides agree on the facts, people may disagree on conclusions because of their different premises." Lind argues that this is why politicians often seem to speak past one another or ascribe different meanings to the same events. In this sense, a worldview can include ideas about nature, values, emotions, and ethics.[50]

For more than 10 years, I witnessed, observed, and analyzed the effects of incompatible worldviews as they played out in the conduct of climate politics. Climate change worldviews can be placed on a spectrum between two opposing positions, which I refer to as *carbon reductionism* and *climate change from the streets*. What follows is a generalization of the dynamics I observed during the initial years (2006–12) of implementation of California's climate change policies. While the perspectives of particular people and organizations were always more complex than these categories may suggest, the categories are nonetheless useful in developing a sense of the sources of misunderstanding, disagreement, and fracture that can hamper public discussion of such policies.

What I refer to as carbon reductionism was broadly based on utilitarianism—efforts to develop climate policy that would deliver the "greatest good for the greatest number." Utilitarianism is an ethical theory that states that the best action is the one that maximizes utility for the greatest number of people in society.[51] In relation to climate change policy, carbon reductionism reflects an adherence to cost-effectiveness and market-based solutions (that is, cap-and-trade) focused on reducing global greenhouse gas emissions. It is argued that although emissions of greenhouse gases and local pollutants may be correlated, greenhouse gas mitigation efforts should not be required to reduce local pollution. These are seen as fundamentally different problems that are best addressed separately.[52] "Carbon reductionism" thus underscores a myopic vision of mitigation; the term literally refers to stakeholders' desire to reduce carbon dioxide emissions (the most prevalent greenhouse gas) while simultaneously emphasizing the reductive logic embedded in this approach. Such assumptions can lead policymakers, businesses, and traditional environmental groups to judge governments according to whether their climate policies are economically advantageous and benefit the majority. While traditional environmentalists use a market logic in hopes of creating climate policy that can be replicated elsewhere, many businesses are in favor of a limited focus because it minimizes government intervention and supports risk management and corporate strategies that lessen the financial burden of mitigation efforts.

Environmental justice groups, in contrast, generally emphasize moral rights, which leads them to a critical reevaluation of both the practice and politics of reducing carbon emissions. The worldview I refer to as climate change from the streets prioritizes equity and justice; from this activist perspective, the utilitarian approach often ignores distinctions between people and the disproportionate impacts of climate change on low-income communities of color. Climate solutions are evaluated on their ability to address environmental disparities and prioritize communities living near polluting sources. Environmental justice groups that support climate change from the streets promote participatory, embodied, and experimental methods in the development of climate policy. They are more willing to consider aggressive policies to reduce greenhouse gases and to transition away from a fossil fuel economy. Carbon reductionism, on the other hand, is focused on incremental, year-by-year greenhouse gas reductions within an existing economic framework. In climate policy debates, environmental justice advocates are apprehensive about market-based solutions because they see them as serving those with wealth and power, rather than the disadvantaged.[53]

I use the categories of carbon reductionism and climate change from the streets to highlight the contentious politics of scale, values, markets, and race in relation to climate change. They help clarify how the actors involved in California's climate policies are often speaking from structural locations that are worlds apart. But I do not represent these categories as unchanging or absolute in practice. Instead, I argue that worldviews can transform over time. Ideas and beliefs about climate change evolve together with the representations, identities, debates, and institutions that give practical effect and meaning to policies.[54]

In the following section, I describe the characteristics that distinguish carbon reductionism and climate change from the streets. Through this synthesis, we can see that responses to climate change are components of larger social, political, and environmental dynamics that combine to shape individuals' ideas about fairness and justice and the role the government should play in enacting solutions. Understanding this larger picture, and examining often unspoken assumptions, allows us to reflect on how common ground can be deliberately negotiated between these positions and, in particular, how environmental justice and public health can meaningfully be integrated into climate policy.

Carbon Reductionism and Climate Change from the Streets

As public concern over the changing climate grows, governments and scientists are focusing more on its chief cause: global greenhouse gas emissions. This results in climate policy with the specific goal of reducing the seven emissions identified by the Kyoto Protocol and California's AB 32.[55] Scientific studies have shown that greenhouse gas emissions have "no direct public health impacts" since they are global pollutants that mix uniformly in the atmosphere. They do not have localized effects like particulate matter and nitrogen oxides.[56] Evidence of observed climate change impacts, moreover, is reported as strongest and most comprehensive for natural systems, a finding that is often used to justify a focus on biophysical rather than social systems.[57]

Although greenhouse gases and pollutants such as particulate matter and nitrogen oxides are often emitted concurrently from processes such as fossil fuel burning, there are challenges to addressing them jointly in climate policies. The most scientifically rigorous and cost-effective method to address climate change is understood to require a strict delineation between global and local scales.[58] Carbon dioxide (CO_2) holds a privileged position in climate policy since it is the most abundant anthropogenic greenhouse gas that contributes to global warming and persists in the atmosphere for many years. To quantify and monitor greenhouse gases, climate analysts convert the gas levels to a "CO_2 equivalent." The CO_2 equivalency of any given gas is calculated by multiplying its mass by the "global warming potential," which indicates the equivalent greenhouse effect of a pound of that gas as compared to a pound of CO_2 (table 1.1).[59]

Such calculations are often linked with a view of nature that focuses on ideas such as biodiversity, ecological integrity, and natural systems.[60] The ambitions of this type of climate change governance are often stated in quantitative terms, such as achieving carbon reduction targets, preventing dangerous anthropogenic interference with the climate system, and limiting the average global surface temperature increase. Given the assumption that climate change should be understood and addressed at the global scale, these ideas provide crucial tools for detecting its progress, measuring its causes and effects, and quantifying the changes to human behavior necessary to

Table 1.1. Greenhouse gas CO_2 equivalents

Greenhouse gas	Global warming potential (20 years)	Global warming potential (100 years)
Carbon dioxide	1	1
Hydrofluorocarbons	100–11,000	100–12,000
Methane	84	28
Nitrogen trifluoride	12,800	16,100
Nitrous oxide	264	265
Perfluorocarbons	5,000–8,000	7,000–11,000
Sulfur hexafluoride	17,500	23,500

Source: Adapted from CARB, *First Update*, 16. Courtesy of the California Air Resources Board

avoid catastrophe for our species and planet. Nonetheless, if used exclusively in public discourse, they can create an abstraction of nature and limit other types of knowledge about the changing climate, including environmental justice perspectives.[61]

These conceptual structures narrow the focus of climate change measures to address greenhouse gas reductions across polluting sources, regardless of place or scale. Since climate change is a global issue and greenhouse gases are global pollutants, it is argued that specific locations for reducing emissions do not matter, as long as reduction targets are achieved. However, if greenhouse gas reductions are coupled with reductions in other associated pollutants, location becomes significant for communities near major polluting sources, such as oil refineries (tables 1.2 and 1.3). As these associated pollutants have direct and immediate health effects, focusing on the distribution of health benefits from local air quality improvements could increase public support for climate change policy.[62]

Other researchers have used another phrase, *carbon fundamentalism*, to explore connections between environmental ethics, climate policy, and economics. Related to market fundamentalism, carbon fundamentalism attempts to capture the conviction of those who believe that market principles should be used to regulate dimensions of social life, including environmental problems.[63] This perceived faith in markets is a defining feature, and it has attracted particular criticism. In popular press articles, climate policy analysts Daphne Wysham and Gopal Dayaneni both use the phrase "carbon fundamentalism" to denote what they see as a misguided belief that market-based systems such as cap-and-trade can provide the greatest equity and prosperity for society. They contend that carbon fundamentalism narrows the framing of climate change, thus obscuring the larger ecological context and the inequitable economic system that created the problem.[64] In the eyes of critics, unrestrained market forces have the potential to exacerbate existing forms of inequity, and an adherence to market fundamentalism may not provide the best solution for every economic or social problem.[65]

Industrial ecologist Braden Allenby and science policy scholar Daniel Sarewitz further suggest that carbon fundamentalist approaches rely on science not for factual enlightenment but as a source of moral authority and a means of implementing contested

Table 1.2. Sources of greenhouse gases and their health effects

Source	Health effects
Carbon dioxide (CO_2) Burning fossil fuels: oil, coal, and natural gas for energy, industry, and transportation Hydrofluorocarbons Refrigeration Air conditioning Solvents Aerosol pollutants Methane Landfills and livestock farming Nitrous oxide Agricultural fertilizers Burning fossil fuels Nitrogen trifluoride Manufacturing of semiconductors, new-generation solar panels, flat-screen televisions, touch-sensitive screens, electronic processors Perfluorocarbons Aluminum smelting Semiconductor manufacturing Substitute for ozone-depleting chemicals Sulfur hexafluoride Car tires Electrical insulation Magnesium industry	Greenhouse gases are global pollutants; they do not have localized health effects.*

* Exposure to greenhouse emissions has human health impacts in concentrated form, such as their use in the workplace (Wisconsin Department of Health Services, "Chemical Fact Sheet").

Source: Adapted from Lin, *Understanding Climate Change*, 7

policy measures. They argue that this form of "green moralism" obscures the complexity of the causes and impacts of climate change. Allenby and Sarewitz write, "The campaign to create a moral universe predicated on carbon footprint, which began with initiatives against sport-utility vehicles, is now extending across society as a whole. Climate change science and climate change policy are rapidly becoming carbon fundamentalism, a simplistic but comprehen-

Table 1.3. Sources of co-pollutants and their health effects

Pollutant / Source	Health effects
Carbon monoxide	
Motor vehicle exhaust	Decreased function of vital organs
Industrial activities (e.g., steel production)	
Ground-level ozone	
Motor vehicle exhaust, oil refining, printing, petrochemicals, lawn mowing, aviation, brush fires	Irritation of the lining of the nose, airways, and lungs Chest pains
Lead	
Mining, metal and cement manufacturing, waste incinerators, battery recycling	Accumulation in bones and teeth Damage to liver, kidneys, and brain
Nitrogen dioxide	
Burning fossil fuels	Lung inflammation and infection
Gasoline and metal refining	Wheezing, coughing, colds, flu, bronchitis
Particulate matter	
Brush fires, dust storms, pollens, sea spray	Respiratory illnesses (e.g., asthma and bronchitis)
Motor vehicle emissions, industrial processes, unpaved roads	Cardiovascular disease, cancer
Sulfur dioxide	
Industrial activity: electricity from coal, oil, or gas that contains sulfur	Coughing, wheezing, shortness of breath
Toxic air pollutants (e.g., benzene)	
Factories, consumer goods, sewage treatment plants	Increased cancer risks
Gasoline additive	Damage to multiple organs and biological systems

Source: Adapted from Lin, *Understanding Climate Change*, 7

sive structure of moral valuation that can be applied to virtually any individual or institution."[66]

"Carbon fundamentalism" is a contested term in the literature. It can carry negative connotations about supporters of market-based climate policies, especially because of its reference to religious zealotry. For this reason, I have adopted the phrase *carbon reductionism* to more neutrally refer to a set of beliefs and practices and to describe the function that they play in larger socio-ecological systems. For example, economists often favor a cap-and-trade mechanism because it harnesses market forces to seek out the least costly carbon mitigation options. However, environmental justice advocates argue that the least costly mitigation solutions do not necessarily maximize equity. In an analysis that goes beyond existing research, I have identified six defining characteristics of carbon reductionism and how it worked in practice during the initial years of implementation of California's climate change policies (2006–12). I have also defined the characteristics of the alternative equity-focused framework, climate change from the streets, which environmental justice groups advanced in response.

Carbon Reductionism

GREENHOUSE GAS REDUCTION POTENTIAL

Carbon reductionism focuses on the greenhouse gas reduction potential of climate policies, which is measured in terms of CO_2 equivalency. For example, California's AB 32 uses greenhouse gas emissions levels from the year 1990 to serve as the 2020 emissions limit: 427 million metric tons of carbon dioxide equivalent emissions. Policymakers contend that Californians' support for such ambitious approaches is motivated by a desire to protect natural systems and the high quality of life they provide. In implementing this law, state officials asserted that California has been "sustained, in more ways than one, by the mountains, deserts, rivers, streams, forests, farmlands, rangelands, coastline, and temperate climate that form [the] natural environment" and define the state. These resources and their "natural beauty," it is argued, together facilitate California's continued economic and cultural growth.[67]

SCIENTIFIC FRAMING

The second characteristic is the requirement that climate change policy be supported by a community of scientific experts. Under such an approach, dispassionate experts advise policymakers of objective truths, after which policymakers factor in social or political considerations. This process implies a linear approach in which scientific facts inform policy, and scientific inquiry is understood to be independent of society, politics, and values. This separation of science from politics results in policymaking that is seen as evidence-based and rational (see Chapter 2 for further discussion on the modes of inquiry and expertise typically involved in this framing).[68]

COST-EFFECTIVENESS

The third characteristic of carbon reductionism is the assertion that appropriate action on climate change requires an array of measures to capture the maximum technologically feasible and "cost-effective" emissions reduction opportunities wherever possible. AB 32 defines the measure of cost-effectiveness as the "cost per unit of reduced emissions of greenhouse gases adjusted for its global warming potential."[69] The concept of cost-effectiveness, as applied to the reduction of carbon emissions, requires policymakers to judge the distribution of costs associated with implementing a climate policy and account for any trade-offs. Attaining cost-effectiveness, however, does not fully allow the specification of where emissions abatement will take place. This raises broad questions: How do different climate policies affect the distribution of costs, benefits, and consequences? And how are these effects experienced across scales and demographic groups?

MARKET-BASED SOLUTIONS

Fourth, cost-effectiveness is often achieved through market-based mechanisms (such as cap-and-trade). This approach is utilized for large industrial emitters, such as electrical generation, manufacturing, cement production, and oil and gas production and supply. According to carbon reductionist logic, incentivizing mitigation and allowing regulated entities flexibility in deciding how best to meet reductions targets spurs market innovation and drives new technologies to higher volumes and lower prices.[70] It is argued that this approach develops market-winning solutions rather than simply mandating emissions reductions at polluting sites (also known as "command-and-control" regulations). Proponents hold that "no other approach can do the job and do it at acceptable cost. By putting a price on carbon, market-based policies harness the power of our free enterprise system to reduce pollution at the lowest costs."[71] Although the cap-and-trade program accounts for less than one-third of the state's total mitigation measures, it remains a central concern for environmental justice groups because the industrial sector's obligations to greenhouse gas reductions are achieved mainly through its compliance under the program (see Chapter 3).[72]

GEOGRAPHIC NEUTRALITY

The fifth characteristic of carbon reductionism focuses on the geographic neutrality of policy interventions. In shaping California's AB 32, policymakers narrowed climate measures to directly address greenhouse gas reductions across polluting sources, regardless of place or scale. Since greenhouse gases are global pollutants that mix uniformly in the atmosphere, it is argued, specific locations for reducing emissions do not matter, as long as California meets its reduction targets. State policymakers, moreover, view California as a member of the global community and envision its climate policies to be part of a larger domestic and international system. The state has strategically chosen to move away from a "parochial" scale—even if attention to local conditions would result in the direct reduction of associated harmful pollutants.[73]

EMPHASIS ON MITIGATION

The last defining characteristic of carbon reductionism is its main focus on mitigation. California's mitigation measures are intended to slow the emissions rate for human-caused greenhouse gases to avoid further disruptions to the Earth's atmosphere. As adopted in 2006, the AB 32 did not initially include adaptation measures. The goal of adaptation measures is to protect lives, health, property, and ecosystems from actual or anticipated climate change impacts, such as heat waves, wildfires, droughts, and flooding. Put another way, mitigation can be viewed as activities that protect "nature from society," whereas adaptation involves ways of "protecting society from nature."[74] Although robust adaptation strategies were developed in later years, early strategies were directed toward protecting hard assets, such as vital infrastructure or ecological systems, not neighborhoods and socially vulnerable populations (see Chapter 4).

Climate Change from the Streets

California environmental justice advocates challenge many aspects of the carbon reductionist worldview. They argue that the framing of climate change as an abstract, global issue excludes the embodied experience of individuals in disadvantaged neighborhoods. Carbon reductionist approaches are seen as separating climate change from political and socioeconomic factors and, most importantly, from the human scale. They can obscure environmental inequalities in communities of color.[75] As described by Hannah Barugh and Dan Glass, this framing privileges experts as the bearers of knowledge about both the problems and the solutions, and one result is that "communities are disempowered to examine the issue for themselves."[76]

California environmental justice advocates are pushing forward an alternative climate policy paradigm. Climate change from the streets challenges the established policy practices of carbon reductionism by embedding justice and public health goals within climate change science and solutions (fig. 1.1). It shifts the focus of policies from the degradation of "nature" to address the simultaneous degradation of communities.[77] Environmental justice advocates know and analyze the effects of climate change through people's histories, culture, and embodied knowledge rather than solely through data gathered by experts and policy implemented by regulatory agencies. In an effort to challenge the prevalence of carbon reductionism in climate policy and action, environmental justice advocates have developed a community-based framework for addressing local and global environmental health impacts. The six characteristics of climate change from the streets, which I have defined, are as follows.

CO-BENEFITS POTENTIAL

Climate change from the streets stresses the need for climate policies that yield substantial health benefits through the reduction of

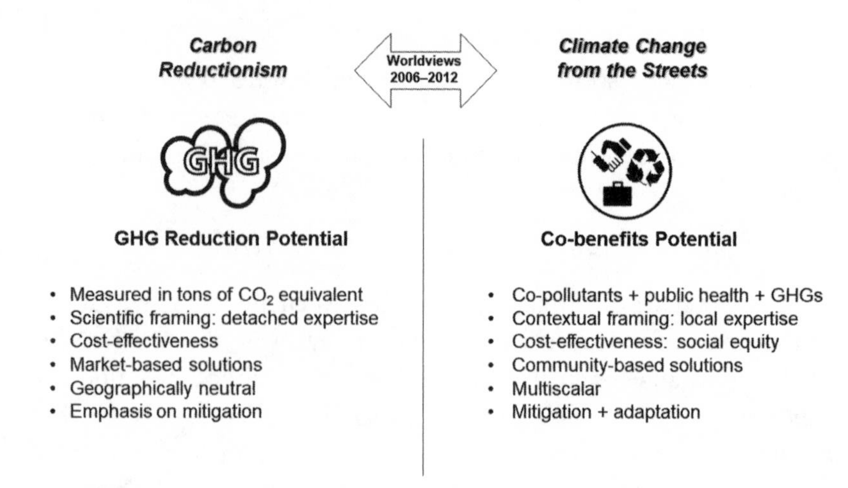

Figure 1.1. Tension between California's climate change worldviews

other pollutants that are commonly emitted alongside carbon dioxide. Environmental justice groups reference valuation studies that suggest that such health benefits may be comparable in magnitude to the value of reduced carbon emissions. They recommend that policymakers should directly compare the cost of climate change actions with the economic value of their benefits, in terms of avoided damage to human health and the environment. For this reason, activists argue that cost-effective and efficient policy design should seek greater emissions reductions where health benefits would be highest and are most needed. Climate policy that ignores such benefits is considered inefficient in two ways. First, it would choose suboptimal emissions reductions targets overall. Second, it would fail to account for differences in abatement benefits across emissions sources (see Chapter 5).[78]

LOCAL EXPERTISE AND EMBODIMENT

Environmental justice groups center their work on issues of embodiment and ask, "What are the connections that their bodies make and manifest daily between the changing climate, pollution, and health?"[79] In this book, I characterize their perspectives and knowledge as focusing on *climate embodiment*, a holistic understanding of the multiple harms that pollution and a changing climate have on human bodies in specific local settings. These types of questions and forms of embodied knowledge supplement solutions developed by policy experts. They confront power dynamics in environmental governance and can alleviate problems that can be caused by introducing one-size-fits-all policy into various local contexts. Environmental justice groups, moreover, debate with experts over issues of truth and method, specifically challenging the political use and control of expertise by claiming to speak credibly as experts in their own right. In other words, activists articulate a form of embodiment that represents a continuum, where the human body cannot be divorced from its environment and where environmental solutions cannot be isolated from the human body.[80]

SOCIAL AND HEALTH EQUITY

Environmental justice groups conceptualize climate change policy in a multifaceted way by calling out uneven impacts while advocating solutions focused on social and health equity. Health equity is achieved when every person has the opportunity to "attain his or her full health potential" and no one is "disadvantaged from achieving this potential because of social position or other socially determined circumstances."[81] In the context of climate change, this approach involves exploring how climate programs, practices, and policies affect the health of individuals, families, and communities. It establishes common goals and ongoing constructive relationships between the health sector, climate science, urban planning, and other fields at multiple scales.

COMMUNITY-BASED SOLUTIONS

In promoting climate change from the streets, environmental justice groups seek to empower individuals and groups with the skills they need to effect change within their communities. They advocate for partnerships with residents, scientists, governments, and other entities as a means to harness policy processes that support community-defined goals. Hence, their solutions to climate change involve diverse stakeholders in the strategic and management activities of climate planning and policy. These approaches, such as transit-oriented development, renewable energy, and urban forestry projects, are aimed at reducing global greenhouse gas emissions and the risk of asthma and respiratory diseases. Advocates also seek to generate career-track jobs in clean technology for workers from disadvantaged communities. These solutions seek to target policy investments and resources to the neighborhoods most in need.

MULTISCALAR CLIMATE POLICY

Climate change from the streets takes environmental justice advocates, whose concerns are often rooted in local conditions, beyond a single site and links them with the multiple scales of climate change. In doing so, it creates opportunities to rethink the relationships among places, people, projects, and sources of knowledge, and it opens up spaces that are often rendered invisible from the point of view of a single location or scale. In this sense, environmental justice activists have rallied behind issues such as air pollution, poverty alleviation, and green jobs to advance equitable climate change solutions at the local, state, and global levels (see Chapter 6).[82]

ADAPTATION AND MITIGATION

The last defining characteristic of climate change from the streets is that its practitioners make concerted efforts toward neighborhood-scale adaptation planning, in addition to mitigation measures. Such integration can be difficult because of important differences in policy objectives. Mitigation deals with the causes of climate change (accumulation of greenhouse gases in the atmosphere), whereas adaptation seeks to prepare society for the impacts of a changing climate. The policies are also defined in different spatial and temporal scales. For example, the benefits of adaptation are local and short-term, whereas mitigation benefits are global and long-term. Nonetheless, environmental justice groups contend that synergies between mitigation and adaptation can be developed in a cost-effective and equitable way. This type of policy integration is important because it acknowledges that certain irreversible and significant impacts from climate change are already underway and are inevitable, even if gov-

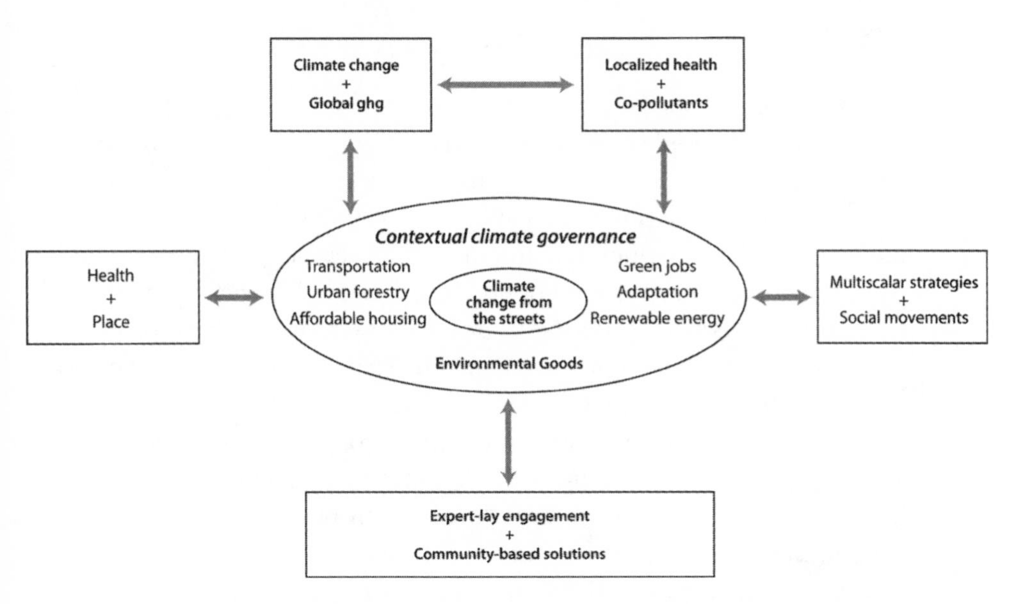

Figure 1.2. Conceptual diagram of climate change from the streets

ernments succeed in reducing greenhouse gas emissions. Environmental justice advocates argue that some groups are more socially vulnerable and will need additional safeguards from the immediate and anticipated effects of climate change.[83]

Environmental justice advocates are demanding a greater role in the decision-making processes that impact their lives. They are debating not only the relationship between political power and climate expertise but also the efficacy of solutions to produce positive social change. Climate change from the streets recognizes reciprocal relationships between nature and society—a perspective that offers a socially and geographically specific understanding of climate change interventions and human health (fig. 1.2). This approach acknowledges how climate change is connected with other types of knowledge about the environment and enables different ways of knowing to play a valid part in framing policy responses. In climate change from the streets, individuals and groups with a range of competing interests work together to develop climate policy and enact environmental justice.[84]

Research Methods

Climate Change from the Streets is a multisited ethnographic policy analysis. I trace people, their conflicts, and their responses to climate change as an open system—one that is continually reconstituted through "the interactions of many scales, variables and forces" to understand what the phenomenon means in related but diverse settings.[85] This approach is based on two key data sources: semistructured interviews with activists, traditional environmentalists, business leaders, and senior government officials; and reflections from my years observing politics in the California State Capitol between 2003 and 2018. The former allowed me to focus my investigation and examine specific questions and themes in greater depth. The latter include participant-observer notes from public hearings and community meetings, conversations with stakeholders, and the textual analysis of climate policies. Together, these materials provide an insider's perspective on the political processes, tensions, and synergies influencing climate change and environmental justice policy in California. In regard to my professional background, from 2003 to 2006, I served as the senior consultant to the California Assembly Select Committee on Environmental Justice. From 2006 to 2010, I held several lobbying positions advocating on health and clinical affairs and natural resources protection; from 2014 to 2016, I served as a gubernatorial appointee to the State Mining and Geology Board (focusing on climate change and land use); and from 2016 to 2018, I was an advisor to a member of the California Air Resources Board.

To understand the narratives that people tell about climate change and environmental justice, I selected interviews by assessing the professional networks I had developed while working in public policy. From that assessment, I conducted 40 semistructured interviews between 2012 and 2015, with 25 follow-up interviews and conversations in 2017 and 2018. My professional work in Sacramento with activists, traditional environmentalists, business leaders, and senior government officials provided a degree of trust and personal connections that resulted in greater access to stakeholders. Many of the people I interviewed remain actively involved in climate policy and planning in California. In order to address sensitive issues, many of the interviewees requested anonymity. In such circumstances, the references indicate only the type of organization that the individual is affiliated with.

The textual analysis of the case studies involved the review of primary and secondary materials and gray literature, such as internal memos and documents from community-based organizations and government agencies. This included archival research on 30 of California's statewide climate policies and analyses of 41 municipal climate action plans and 10 regional climate and sustainability plans. The textual analysis also relied on newspapers and popular press articles, as many of the climate action initiatives discussed here are fairly recent. Throughout, I sought to identify evidence of the development of climate policies that substantively engaged participants outside government, in order to establish explicit interconnections among climate change, environmental justice, and public health.[86]

Through these methods and sources, I identified people and organizations, concepts, and critical sites of tension in California's climate policymaking process. I centered my multiscalar analysis on key case studies that outlined the dynamics of conflict and collaboration. Environmental justice advocates working on local climate action in the cities of Oakland and Richmond were also simultaneously lobbying in the halls of the state capitol and building translocal coalitions with Indigenous rights groups in Mexico and Brazil. Hence, the case studies presented in this book have overlapping temporal and spatial scales. The narratives of environmental justice activists are *not* told in a linear fashion. Their movements across space and time are analyzed as overlapping nodes in multiscalar networks rather than discrete actions within geographical boundaries drawn at a specific scale (see the timeline in the Appendix).[87]

In advancing this work, I tell the stories through which the environmental justice movement defines climate change. Their stories of conflict and collaboration provide an opportunity to reflect on the strengths and weaknesses of particular environmental protection paradigms and the prospect to imagine politics differently. Such stories, moreover, have the power to transform meaning and shift day-to-day realities. I hope they will inspire leadership and facilitate coalition-building within California and beyond. The choices that are being made about climate change today will affect the well-being of generations to come. It is crucial that all concerned, including policymakers, activists, scholars, urban planners, and community members, approach these decisions with a true sense of the possibilities open to them. Climate policy has the potential to address not only the global threat of greenhouse gas emissions but also the patterns of racism and inequality that expose specific communities and individual bodies and lives to pollution's gravest dangers. If it does not achieve both, it will likely be destined to fail. The stories that I tell in this book suggest a way forward.

CHAPTER TWO

Climate Embodiment

At Peres Elementary School in the city of Richmond, California, the students practice not only emergency earthquake drills but also chemical explosion drills. The public school is situated less than a mile away from a major oil refinery. When an alarm warns of an incident at the refinery, teachers run to shut doors and close windows. Students are told to cover their faces with tissue paper. In the surrounding community, residents are given "shelter in place" warnings. They are instructed to stay inside, close windows and doors, turn off air conditioning and heating units, and use duct tape to seal off airways. Children growing up next to the refinery often report problems such as asthma, headaches, and other ailments. According to long-term Richmond resident Sandy Saeteurn, "There were incidents at the refinery when I was little, when they would have a spill. . . . I would get so sick that I would have to go to the hospital. My eyes swelled almost closed, and my body would be covered in hives."[1]

For communities living next to polluting industries, preparing for emergency explosions and seeking medical care for the short- and long-term consequences of breathing some of the nation's dirtiest air are fundamental to their experience of the local environment. Throughout California, these sources of pollution are disproportionately located in low-income communities of color, which have far fewer resources to resist politically, adapt, and recover. Studies show that environmental regulations in these communities are poorly enforced. In Richmond, violations of air quality rules are frequent. According to city officials, over a 10-year period, there were 13.1 violations per 100,000 people, compared with 0.96 for the entire San Francisco Bay Area.[2] Environmental health risks raise a central question for these communities: What are the links that their "human bodies make and manifest every day between [inequities], disease, and death," on the one hand, and environmental justice and well-being, on the other? Or framed another way: How do the dynamics of "power and inequity express themselves through population distributions of health, body size, disease, disability, and death?"[3]

Areas such as Richmond have given rise to forms of activism centered on environmental justice, which attempt to redress the structural and historical inequities that expose low-income communities of color to harm from fossil fuel burning and other forms of industrial pollution. The ways in which environmental justice advocates view climate change are directly informed by their lives in such communities. Many have experienced or seen firsthand the physical harm that residents suffer in neighborhoods close to polluting facilities. Their perspectives on pollution are embodied and concerned with its physical and social dimensions. Their measures of pollution are often community-based and qualitative rather than abstract and quantitative. Their local values and goals tend to be at odds with those of the policymakers, traditional environmental organizations, scientists, and bureaucrats who shape climate action at a global scale.[4]

This chapter examines the theories and perspectives that environmental justice advocates bring to their work on climate change from the streets. Key to their worldview is an understanding of the human body as a site of intersection between social, political, and environmental dynamics. Activists view climate change as an embodied phenomenon that has multiple impacts on the people who live with it every day. These insights are tied to a keen sense of the ways in which health is determined not just by personal choices or genetic makeup but by a range of social and historical factors that hinge on race and class. Through organizing and lobbying efforts, environmental justice advocates have been able to disrupt the dominance of the carbon reductionist worldview in climate policy. They have introduced embodied, local forms of knowledge and perspectives into public debate and transformed climate change solutions. The worldview that environmental justice advocates share has its roots in communities such as Richmond, and its origins and workings are key to understanding climate change from the streets.

Richmond is home to one of the world's largest oil companies, the Chevron Corporation. Its refinery is also California's single largest source of greenhouse gas emissions. Chevron released 4.5 million metric tons in 2010 alone. Established over a century ago, it processes more than 250,000 barrels of crude oil each day. Chevron sprawls across five square miles of Richmond, and its smokestacks are as tall as skyscrapers. The hundreds of tanks that dot the refinery can hold up to 15 million barrels of gasoline, jet fuel, and diesel fuel. The Chevron refinery releases nearly 600,000 pounds of other hazardous and toxic emissions into Richmond's environment each year.[5]

Over 25,000 people, including those in two public housing projects, live within three miles of the refinery, and less than a mile away are playgrounds and two public schools. Nearly 85 percent of these residents live below the federal poverty level, and more than two-thirds are people of color. Environmental justice groups argue that refinery pollution not only fuels climate change but also contributes to high rates of asthma, cancer, and heart disease. According to the California Department of Public Health, Richmond residents of all ages are almost twice as likely to go to hospital emergency rooms for asthma attacks as others living in the region. African Americans in the area, in particular, have the highest rate of asthma emergency visits and admissions in the county—at a level four times than that of other racial groups. A 2018 study by Lara Cushing et al. found that California facilities emitting the highest levels of both greenhouse gases and particulate matter tend to be located in neighborhoods with higher proportions of residents of color. Another study has confirmed this trend—even when controlling for household income—for a wider range of co-pollutants, including particulate matter, nitrogen oxides, and sulfur dioxide.[6]

Chevron's Richmond refinery exemplifies the difference between mainstream and environmental justice views of climate change and its relationship with various forms of pollution. The chief objective of California's climate policy is to reduce global greenhouse gas emissions. But environmental justice advocates argue that climate action should also address equally pressing problems with local air quality, which are caused not by greenhouse gases but by other harmful substances that are often released with them into the atmosphere through the burning of fossil fuels. Benefits from the direct reduction of greenhouse gas emissions can yield important opportunities for public health in the form of reduced co-pollutant emissions.

Environmental justice groups begin with the premise that climate change policy should take a holistic approach that integrates multiple pollution dimensions. For them, efficient policy design would seek greater greenhouse gas emissions reductions where health co-benefits are the most needed. Activists contend that the failure to maximize pollution reductions in the state's most affected areas, such as Richmond, could reinforce a history of unequal exposure. Several environmental health researchers have supported this perspective and argue that policies that focus exclusively on greenhouse gases can be ineffective in addressing pollution, potentially reinforcing structures of environmental inequality.[7]

Climate change is a complex problem involving intricate interactions among biological, physical, and social systems. Holistic solutions demand collaboration among scientists, policymakers, and diverse publics, including active consultation with the communities most burdened by climate impacts.[8] Using the concept of climate embodiment, I examine how environmental justice groups are creating new forms of environmental expertise. This expertise has its roots in experience. Residents of neighborhoods such as Richmond are literally embodying climate impacts and pollution and using their lived experience to enact social change.

Lived Experience and Bodily Impact

Humanizing climate change is not an easy endeavor. Policymakers and traditional environmental organizations are often invested in global-scale perspectives of the problem. Pinpointing its effects at the community and individual scale can be difficult, in part because the science of how people embody multiple impacts is complex and can be contentious. Environmental justice advocates, however, are gaining influence in the policymaking process. Their work focuses on the harms that pollution and other industrial activities do to human bodies. According to the gender studies scholar Jade Sasser, for advocates, climate change, like all environmental change, is an embodied experience. It is a living phenomenon, in which "shifts in temperature, weather patterns, atmospheric concentrations of gases, and policy negotiations act on, with and through human bodies in a variety of complex ways that are seen and unseen." Through changing respiratory and infectious disease patterns; increased deaths from heat waves, wildfires, and extreme weather events; and sanitation and hygiene problems from shortages of safe drinking water, we can witness the bodily impacts of this dynamic phenomenon.[9]

I was introduced to an embodied approach to climate change when I met Mari Rose Taruc, an advocate with the Asian Pacific Environmental Network. Taruc's experience with this group led her to fight for formalized, state-level assessments of the impacts that pollutants from multiple sources can have on the bodies and well-being of local residents. This multifaceted, experiential approach has directly shaped her activism around climate change. I sat down with her a year after a major fire at the Chevron refinery in Richmond. The fire on August 6, 2012, engulfed 19 workers and sent 15,000 area residents fleeing to hospital emergency rooms after suffering from respiratory problems and vomiting (fig. 2.1). It created a vapor cloud, over four miles wide, of sulfuric acid and nitrogen dioxide—chemicals that can affect the skin, eyes, and respiratory and gastrointestinal tracts. The fire also released greenhouse gas emissions that fuel climate change. This was the third major fire at the oil refinery in 12 years. All were caused by leaking pipes, and federal investigators blamed the fires on Chevron's negligent safety culture, flawed emergency response, and failure to inspect and upgrade vulnerable piping.[10]

Taruc, a Filipina American whose Indigenous Tagalog last name means "to know," has worked for over 15 years in the environmental justice movement. She has traveled to the United Nations climate negotiations and been a key leader in statewide coalitions. Since 2003, her organization has fought the expansion of the Chevron refinery, defended the state's greenhouse gas reduction law, established multilingual refinery emergency warning systems, and pushed

Figure 2.1. Commuters step out of their cars to take pictures of the fire raging within the Richmond Chevron oil refinery on August 6, 2012. Courtesy of Communities for a Better Environment

for sustainable development in low-income communities of color. However, she told me that her activism is deeply connected to the fact that she is a mother of two sons who have asthma: "It's mama being protective of her children—but it is also linked to a broader understanding of what it means to protect my children and other kids just like them from sources of pollution that are making them sick."[11] Taruc's activism acknowledges the ways in which "race, class, gender, and geography" structure how bodies and environments are entangled with one another and demonstrates how environmental justice groups reimagine climate change in the context of embodied experience. She is a reminder that environmental knowledge can be established within bodies and therefore shaped by viewpoints that are partial and situated.[12]

When I spoke to Taruc about the 2012 Chevron fire, she described the dark vapor clouds and the smell of toxic ash that was falling onto people, homes, and community gardens. Members of the Asian Pacific Environmental Network worked throughout the night of the fire to assess whether their community was safe: "We could not locate some people. The people we did find were either throwing up or getting rashes. And the new emergency warning system did not fully work. It was a real disaster."[13] The multilingual emergency warning system, developed after a 1999 refinery fire, reached only a fraction of the 20,000 residents who were at risk. Many residents waited three hours for an emergency telephone call that advised them of "shelter in place" safety instructions.[14]

The next morning, instead of joining her colleagues in Richmond, Taruc traveled to the state capitol to speak with regulators at a meeting concerned with the cumulative effects of pollution. To her, it was an important opportunity to "make real" the need for assessment tools that safeguard communities from multiple pollution sources. Since 2003, in light of federal inaction, environmental justice groups had been engaging the California Office of Environmental Health Hazard Assessment over this issue and precautionary approaches to environmental protection.[15] After years of deliberation, in 2007 regulators agreed to establish a work group to develop scientific methods for measuring the cumulative effects of pollution at the neighborhood level. To guide this experimental process, the group adopted two key terms:

> *Cumulative impacts* means exposures, public health, or environmental effects from the combined emissions and discharges in a geographic area, including environmental pollution from all sources, whether single or multi-media, routinely, accidentally, or otherwise released. Impacts will take into account sensitive populations and socio-economic factors, where applicable and to the extent data are available.
>
> *Precautionary approach* means taking anticipatory action to protect public health or the environment if a reasonable threat of serious harm exists based upon the best available science and other relevant information, even if absolute and undisputed scientific evidence is not available to assess the exact nature and extent of risk.[16]

At work group meetings, environmental justice advocates argued for assessing the combined health effects of various pollutants rather than considering them one at a time. They pointed to numerous studies that show that multiple pollution sources are disproportionately concentrated in low-income communities of color. They also cited studies that reported increased sensitivity to pollution for communities with low income levels, low educational attainment, and other biological and social factors. The combination of multiple pollutants and increased sensitivity in these communities, advocates claimed, could result in higher cumulative health impacts.[17]

In debates over methods to address complex and uncertain risks, environmental justice advocates invoked the precautionary principle, framed through two main components: "preventive action in the face of uncertainty and reversing the burden of proof."[18] Advocates saw this principle as a paradigm shift. The status quo, they contend, requires environmental justice groups to prove with scientific certainty that harm has occurred before remedial action can be required. Under the precautionary principle, when reasonable suspicion of harm is raised, the burden of proof is instead placed on government and industry to prove that preventive action is unnecessary. The precautionary approach thus encourages the evaluation of alternatives for the purpose of preventing or minimizing harm.[19]

Such a shift would require a strong investment in research and tools and a government commitment to form a new paradigm of environmental protection. By 2012, the day after the Chevron fire, state regulators still had not adopted a scientific methodology to assess cumulative impacts at the neighborhood scale. The precautionary approach was controversial and opposed by business groups. Over the preceding two years, the Western States Petroleum Association and the American Chemistry Council had argued that the studies referenced by advocates provided "a superficial view of the literature, without critical assessment of study strengths and weaknesses. Contrary/inconsistent evidence is either ignored or downplayed, . . . present[ing] a one-sided assessment of the literature, apparently intended to persuade rather than inform."[20] Representatives of polluting industries worried that the adoption of a cumulative impact methodology could prompt additional environmental reviews and health assessments beyond the scope of existing regulations.[21]

At the cumulative impact work group meeting on August 7, 2012, over 50 stakeholders were present, and emotions were running high. As the meeting began, regulators asked participants to introduce themselves. Taruc seized the opportunity to underscore the urgency for adopting cumulative impact assessment tools: "My name is Mari Rose Taruc, and I work for the Asian Pacific Environmental Network in Richmond. And there is a fire going on right now, where thousands of people are going to the hospital." She then looked over at representatives from the oil industry and said, "Because your refineries are exploding and due to your gross negligence to our community. This is the reason why we need cumulative impact assessment and mapping tools to take care of communities that are suffering from dangerous industrial facilities that are still so rampant out there."[22]

Taruc went on to argue that cumulative impact analysis provides a fuller picture by examining multiple chemicals, various pollution sources, and the characteristics of the population that influence health outcomes. She asserted that assessing and mitigating cumulative impacts were logical next steps in applying the best available science to environmental protection programs, including those related to climate change. In the years since California passed its landmark climate change law, AB 32, activists have warned that the state's trailblazing efforts to reduce greenhouse gas emissions, in particular the cap-and-trade program, ignores the needs of communities of color that are already burdened by pollution from multiple sources.[23]

Taking a long-term perspective, environmental justice groups argue that the effects of the harmful local pollutants that are often emitted with global greenhouse gases during fossil fuel combustion should be a key consideration in climate policy because such policies will influence important features of society, including energy systems, infrastructure, industry, and community design.[24] Their position is based on the relationship between greenhouse gas emissions and fossil fuel use and the idea that addressing greenhouse gas emissions is about fundamentally changing the way we make and use energy. As California's Environmental Justice Advisory Committee has emphasized, "People of color and low-income communities are being crushed under an impossible load of emissions from fossil fuel usage."[25] For them, it is clear that climate change policies that regulate fossil fuel usage can also have the potential to improve the environmental health of these communities. By framing climate change and its effects on human bodies through such connections, environmental justice advocates are taking an experimental approach that promises to yield new insights but also faces the challenge of gaining credibility in relation to established ways of producing and evaluating knowledge.

Embodied Carbon versus Climate Embodiment

At the cumulative impacts meeting and subsequent climate policy convenings, state regulators showed little interest in investigating how greenhouse gases and related pollutants are interlinked with respect to sources, atmospheric processes, and environmental health effects. Instead, they were moving forward with single-purpose assessment tools to reduce greenhouse gas emissions within specific economic sectors, such as in the building and construction industry. The assessment of "embodied carbon" was characteristic of their siloed policy approach.[26]

Embodied carbon, as a concept, recognizes that activities such as mining, driving trucks, running factories, and refining chemicals to produce building materials and products result in greenhouse gas emissions: "Embodied carbon is the sum impact of all the greenhouse gas emissions attributed to the materials throughout their life cycle—extracting from the ground, manufacturing, construction, maintenance and end of life/disposal."[27] Embodied carbon assessments can be used to identify carbon-intensive building materials, designs, and construction processes and to promote alternatives that reduce the amount of carbon released. Embodied carbon, however, does not consider other pollutants or how the extraction, processing, and transport of raw materials affect the health of people within local communities.[28]

Moving beyond the notion of embodied carbon, the environmental justice movement has long pushed for an approach to climate change policy that focuses on the harms that multiple pollution sources have on human bodies, or what I call *climate embodiment*. This term refers to the multiple knowledges and the human impacts of climate change, as well as how these impacts are facilitated through the diffuse infrastructural body (for example, oil refining, extraction of raw materials, transportation, construction of buildings). Environmental justice advocates contend that their bodies tell stories and bear the marks of environmental interactions. Instead of focusing on how buildings embody carbon emissions, they center their

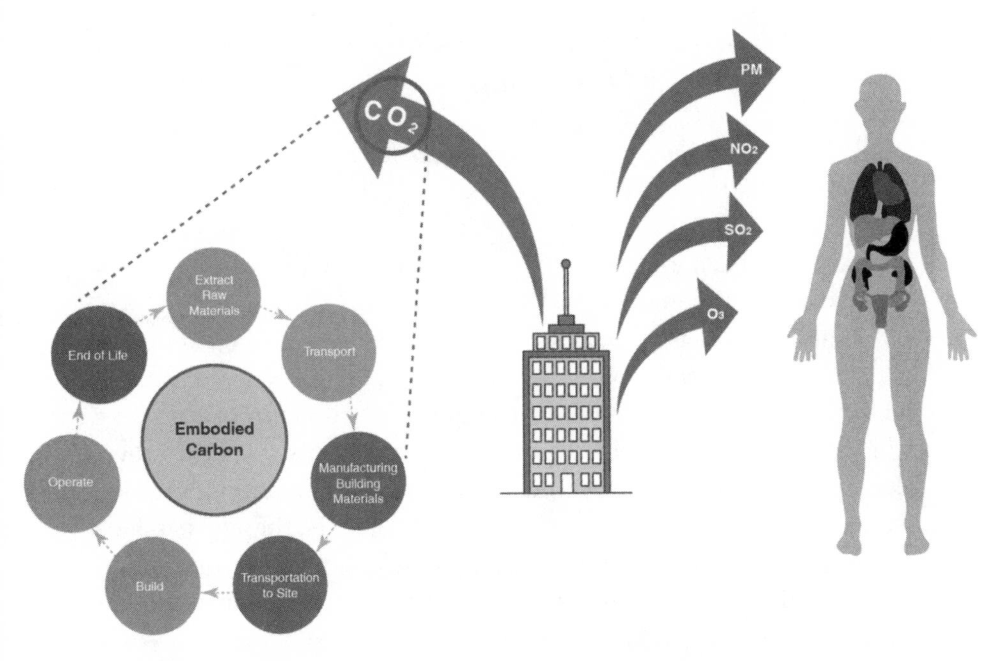

Figure 2.2. Climate embodiment: pollution in the infrastructural body and the human body

work on the human embodiment of climate change and the harmful pollutants that may be associated with carbon emissions (fig. 2.2).

Climate embodiment acknowledges how environmental health is the result of various interrelated social and ecological systems and processes. Bodies become places where points of pollution, inequality, and poverty intersect. Environmental justice advocates challenge assumptions in scientific practice that discount how people embody pollution, and they argue that regulators are often not well equipped to draw holistic linkages from the infrastructural body to the contaminated human body.[29]

For example, in Richmond, environmental justice advocates ask how residents embody the cumulative impact of pollution from industrial facilities. While the Chevron refinery is the state's largest greenhouse gas polluter, people in Richmond also live near four other oil refineries, three chemical companies, eight federally designated Superfund sites, and dozens of other toxic waste sites. In particular, they contend that years of exposure to toxic air emissions and lung-penetrating particulate matter from industries, truck routes, and rail lines adjacent to neighborhoods may also be affecting residents. However, the consequences of these conditions have not been well studied. Epidemiologists often find it challenging to prove a direct correlation between exposures and disease because of confounding factors such as smoking and diet. Conventional health studies have been unable to fully determine why the asthma rate is so elevated in Richmond.[30]

These questions resonate with *ecosocial* theory, a conceptual framework developed by the public health scholar Nancy Krieger. Ecosocial theory articulates connections between activism in practice, on the one hand, and the relevant theoretical perspectives, on the other. Krieger has expressed how individuals biologically embody exposures arising from societal and ecological contexts, thereby producing varying population health outcomes.[31] She argues that while every health condition has a biological basis, the vast majority of health outcomes are facilitated or exacerbated by social and environmental conditions, such as education, income, ethnicity, and geographical location. Ecosocial theory addresses the question of who and what drives social inequalities in health. The "eco" in "ecosocial" refers to "notions of ecology, recognizing that we simultaneously are but one biological species among many. . . . The 'social' in turn draws attention to features of the social realm—created by humans . . . that shape, if not drive, inequitable population patterns of disease, disability and death." The goal of the theory is to integrate social and biological reasoning to produce new insights and possibilities for addressing the causes of disease and factors that determine the distribution of health and illness across a given population.[32]

Krieger focuses on how socially produced pathways of disease, mediated by physiology, behavior, and gene expression, can affect the development, growth, regulation, and death of a person's biological systems. This is in contrast is to conventional frameworks that treat the causes of disease and of group differences in disease rates as primarily biological. As a theoretical framework, ecosocial theory can help health practitioners put social systems at the center of their thinking about how to gather and interpret relevant information about the causes of disease.[33]

Ecosocial theory enables analyses of the exposure and susceptibility of communities. This includes how people "resist injustice and its health-harming effects, individually and collectively, and the resilience that enables them to do so." Ecosocial theory further requires an explicit consideration of how types and levels of exposure, spatiotemporal scales, and historical context shape bodies and embodied experience. In sum, Krieger argues that our bodies are both social and biological because, as the concept of embodiment implies, what we manifest in our bodies, how we ail, and the type of medical therapies we have access to are simultaneously articulations of our experiences in the world and their literal assimilation within us.[34]

Environmental justice groups agree that many of the challenges in Richmond can be attributed to historical context—especially discriminatory land-use decisions. The challenges that their communities face are the same ones that gave rise to the US environmental justice movement in the 1980s. In the book *To Place Our Deeds*, historian Shirley Ann Wilson Moore documents how land-use practices placed a majority of polluting industries in Richmond's communities of color. She found that shortly after Richmond's incorporation in 1905, the city turned to segregated housing. African Americans arriving from southern states and seeking jobs in shipyards were forced to concentrate in areas in North Richmond. Moore describes North Richmond at that time as a community with chemical companies, a landfill, and an oil refinery. The area had limited city infrastructure and services such as streetlights, paved streets, and fire and police protection.[35] According to Betty Reid Soskin, a lecturer on the African American experience during World War II at the Rosie the Riveter/World War II Home Front National Historical Park in Richmond, "It was the only land available to them when they wanted to purchase property. People don't put themselves in harm's way intentionally. Real estate developers could determine where you lived. The local bankers could determine who could get mortgages. Social policy determines history. We have developed sensitivities to environmental injustice, and those sensitivities did not exist during that time."[36]

The insights of environmental justice activists are confirmed by academic research that takes multidisciplinary approaches to examine the interaction of bodies with their multilayered contexts. For example, embodied sensitivities to environmental injustice can also be analyzed through the concept of the *social determinants of health*:

the economic, physical, and social conditions in the environments in which people are born, live, learn, work, and play that influence a variety of health outcomes and risks. Research on the social determinants of health examines the role that culture, gender, race, political economy, land-use patterns, and other social factors have in making human populations vulnerable to disease and poor health. Resources such as affordable housing, education, job access, public safety, and neighborhoods free of toxic substances can enhance quality of life and can have a significant influence on the health of specific segments of the population. Residents in low-income neighborhoods are more likely to live in substandard housing that can expose children to hazards such as lead poisoning and poor drinking water. Children from low-income neighborhoods are also more likely to attend underachieving schools and have less access to quality jobs, which can limit social mobility and health across generations.[37]

The domain of human health, moreover, can be seen as the result of spatial processes shaped by interactions between social and ecological systems. Health outcomes can differ according to disease type and the ways in which built environments can increase individuals' vulnerability to poor health, such as dilapidated housing and proximity to polluting facilities. According to the medical geographer Brian King, the domain of health is fluid because it can expand or contract over time, depending on the current state of social and ecological systems and whether they are undergoing change. Change can occur through state intervention, ecological disturbances, or development projects that modify people's vulnerability. These factors, according to King, constitute the political environmental context.[38]

The political environmental context influences the ways in which particular diseases are understood and managed while also reinforcing limits on health decision-making. It determines which individuals are healthy or able to manage illness through differential access to facilities, environmental amenities (such as parks and recreational spaces), technologies, and medical care. For example, King contends that while disease distribution can be examined "as the product of a natural event that disrupts water and sanitation systems, the ways in which social systems are organized has proved critical in mitigating the severity of an outbreak." Furthermore, human health can change over time within social systems (for instance, land-use patterns, political economy, and institutions) that are destabilized by climate change impacts such as drought, wildfires, extreme weather events, and flooding. Attention to the social determinants of health exposes how people embody climate change impacts in several ways: First, it recognizes the historical and geographical unevenness of health impacts. Second, it illustrates how resources and environments shape health outcomes. Third, it examines the power asymmetries that influence health inequities. And fourth, it focuses investigation on multiple systems and scales that produce unfair and avoidable differences in health.[39]

The medical anthropologists Nancy Scheper-Hughes and Margaret Lock promote a similar approach. They define the body through three perspectives: the individual body, the social body, and the body politic. The individual body is understood through the "lived experiences of the body-self"; the social body is "a natural symbol with which to think about nature, society, and culture"; and the body politic refers to the "regulation, surveillance, and control of bodies (individual and collective) in reproduction and sexuality, in work and leisure, in sickness and other forms of deviance and human difference." Scheper-Hughes and Lock stress that bodies not only are biological entities but also have a social basis that can exhibit signs of marginalization and injustice. They argue that an analysis of embodiment should "begin from an assumption of the body as simultaneously a physical and symbolic artifact, as both naturally and culturally produced, and as securely anchored in a particular historical moment."[40]

Embodied Knowledge and Policy Ruptures

Despite the direct and indirect threats to human health that greenhouse gases and their co-pollutants hold, as well as the complex social processes that shape and channel those threats toward specific communities, an embodied approach to climate policy is rarely taken. There is growing evidence that policymakers are neglecting the disproportionate impact climate change will have on low-income neighborhoods and communities of color.[41] This is in part because of the kinds and sources of knowledge considered to be valid within the frame of public decision-making on climate change. Scientific knowledge tends to carry the greatest authority, while the embodied knowledge developed within local communities is often dismissed as anecdotal.

Scholars from the diverse fields of human geography, urban planning, critical policy studies, public health, and science and technology studies have found that experts advising state and local governments consider successful climate policy to be grounded primarily in scientific assessments. These experts often underemphasize the socioeconomic conditions in which the phenomenon takes place.[42] For example, JoAnn Carmin and David Dodman conducted focus groups with leading practitioners in 14 cities in North America, Europe, Asia, and Africa to understand what constitutes "successful" uses of science and management of scientific uncertainty in urban adaptation planning. They concluded that scientific evidence plays a critical role in local climate policy. Climate leaders use science "as a discursive and symbolic tool" to legitimize climate-related activities and provide practitioners with the means to set priorities and identify appropriate cost-effective planning measures.[43]

The emphasis placed on scientific assessments as a foundation for climate planning is aligned with the positivist assumption that empirical data and scientific methods are the only sound basis of knowledge and hence will guide policymakers to adopt the best course of action. Carmin and Dodman's study of local climate practitioners identified three critical limitations to a strictly positivist perspective in climate policy: First, it ignores the fact that scientific analyses and findings are themselves the product of social and political processes. Participants' knowledge base shapes both the types of assessments conducted and their findings. Second, though science is often viewed as producing unbiased and irrefutable truths, findings are often used to advance political agendas. When results diverge from the preferences of policymakers, they may be compelled to change their options or challenge the methods or findings. And third, "while many scholars and practitioners are focusing on how to design assessment processes that cities can adopt, as is the case with science in general, the estimation used in these and other types of urban climate assessments are inherently social and uncertain."[44]

Other scholars argue that positivist approaches in local climate policy do not always result in substantive or swift climate action. In many situations, the more that city officials understand about climate change and its consequences, the more difficult decision-making can become. On one hand, climate adaptation theory suggests that as uncertainty increases, a stronger emphasis is placed on flexibility, iterative progress, reflection, and continual learning as information becomes available. Yet on the other hand, under conditions of uncertainty, local policymakers are called upon to be proactive in defining scientifically sound models of climate governance while weighing competing economic and political interests.[45] This results in "a tension between carefully engaging the science so as to create policy that will cope with uncertainty on the one hand, and the call for bold and transformative leadership on the other. It is unsurprising, then, that some city-scale decision-makers find themselves uneasy about taking climate change decisions in a systematic and responsible manner. The fact that climate change is only one of the competing imperatives—alongside critical issues . . . such as economic growth—demanding the attention of city officials and politicians makes defining the right action even more contested."[46]

Local governments often minimize the value judgments underlying climate risk assessments and claim that science primarily influences their decisions. According to Joyce Rosenthal and Dana Brechwald, technical rationality has guided many climate-planning decisions in major cities across the United States, a trend that has the potential to reinforce existing patterns of privilege and exclusion. They assert that a strong reliance on technical rationality has often resulted in the overconcentration of noxious uses in low-income neighborhoods and communities of color that have historically lacked the capacity to resist such facilities. These communities are often marginalized in public decision-making processes and less likely to engage in the technical analyses that are required in regulatory processes. In the United States, urban and environmental planning has long depended on risk assessment and cost-benefit methodologies to meet legal requirements for environmental review of development projects and to ensure local economic development priorities.[47] Urban planners often reinforce positivist approaches by structuring public consultation processes in ways that exclude disenfranchised groups from meaningful participation. This includes limiting the distribution of resources and technical assistance to disadvantaged groups even when policy issues being considered have strong equity implications.[48]

In keeping with these trends, the paradigm of carbon reductionism often discourages meaningful public engagement and the importance of local context. However, the questions of who counts as an expert, whose knowledge is accepted, and who participates in advisory and negotiating bodies are critically important in the development of climate governance. In addressing environmental challenges, strict boundaries between scientific and lay expertise can often be counterproductive. They fail to acknowledge how developments in climate knowledge are linked to historical experience, social practices, or experimentation from the streets. According to Sheila Jasanoff and Marybeth Martello, "To discover new facts about 'nature,' we change ourselves. To build on 'natural' facts without taking stock of associated social orders is risky business."[49] Given the major blind spots in the positivist approach to climate policy, diverse ways of knowing can provide essential safeguards against the continuation of unjust social, political, and environmental structures. Embodied approaches provide a working model for integrating diverse forms of knowledge and encouraging experts to collaborate and develop findings in new settings and at multiple scales.

More research is needed to understand how the environmental justice movement has opened up spaces for the transformation of environmental policy. Approaches such as climate embodiment can be seen as important sources of "ruptures" in technical practice: opportunities that can disrupt the apparently seamless legitimacy of expert-driven policy. Regulatory systems rely on experts to set and define standards based on technical norms produced in labs—and outside affected communities. As a result, the knowledge that community groups bring to environmental problems is often disregarded or deemed irrelevant in the context of environmental governance and technical practice.[50] Yaron Ezrahi contends that by privileging positivist methods, experts and governments often retain legitimacy and power in science and environmental policy decisions. Those in authority are able to describe their views as objective and "technically disciplined" in contrast to the purported emotional and biased positions of lay citizens.[51] Such attitudes can create profound inequities that undermine democratic participation. Policymakers often do not mitigate the effects of pollution on a community until that pollution is proven to have directly caused residents' health issues. Such proof can be an insurmountable hurdle for environmental justice groups, given the scientific standards that govern it.

Through an embodied approach, we can begin to understand how environmental justice groups challenge and transform the knowledge practices of the regulatory institutions governing climate change. Environmental health scholars contend that local configurations of climate change policies and scientific knowledge will depend on the analysis of both the structural and cultural aspects of the civic environments in which decisions are made. Expanding participation in decision-making will also require a broader understanding of the context in which lay people have constructed climate knowledge and expertise and of how this context can vary within and across communities. Such an approach "highlights the importance of considering the intersection of race, class, and gender, the actions of environmental justice social movements globally and within communities" when developing climate change policies.[52]

In *Street Science*, Jason Corburn underscores the particular ways in which local and professional knowledges diverge in the environmental policymaking process. He analyzes local knowledge, a component of embodiment, by asking a series of questions about its production.[53] First, he asks who holds local knowledge; where does it tend to emerge, both institutionally and culturally? Second, he asks how knowledge is acquired. Local knowledge is not mere belief but is subjected to "at least common-sense tests of logic, coherence, and rationality that fit with larger community understandings." Third, Corburn asks what makes evidence credible. People often evaluate the credibility of knowledge by whether or not they participated in its production. In the case of environmental hazards, this may mean community members participating in testing and sampling procedures. A fourth question considers the forums in which knowledge is legitimated. Corburn acknowledges here that in some public forums, "both local and professional knowledge may be competing for legitimacy, such as in the media, so the distinctions between local and professional knowledge may be constantly renegotiated." Finally, he asks what orientation each type of knowledge has toward policy action. Through this analysis, Corburn outlines the ways in which policymakers are beginning to incorporate the expertise of professionals with the contextual intelligence of lay publics. However, he concludes that local knowledge is not a panacea. Rather, it offers a policy framework for "developing polycentric, interactive, and multiparty processes of knowledge-making within institutions that have worked for decades to keep professional expert knowledge away from the vulgarities of populism and politics."[54]

Other scholars have pointed out that in his discussion of professional knowledge, Corburn conflates two types of science: research and regulatory. Research science aims at extending knowledge and competence in a particular area without regard for practical application, while regulatory science aims at improving existing practices, techniques, and processes to further the task of policy development. This includes knowledge production, synthesis (for example, evaluation, screening, and meta-analysis), and prediction of future risks or costs.[55] Raoul Liévanos et al. argue that it is necessary to further elaborate on the conceptual distinctions between regulatory, research, and street science (that is, embodied and local knowledge) when attempting to explain how activists have transformed each: "Understanding the ways in which scientific modes at once influence each other and resist influence can help inform innovations in environmental justice scholarship and practice." These three spheres of science, modes of inquiry, and expertise are typically involved in movements for environmental justice (table 2.1). It is argued that regulatory science, in particular, is influenced by multiple factors. Political, economic, moral, and emotive dynamics shape the production of policy-relevant knowledge in ways that are obscured by the focus on mere professional knowledge initially described by Corburn.[56]

A particularly powerful example of how embodied knowledge can be brought into dialogue with different spheres of scientific authority is in work on "embodied health movements." This scholarship analyzes how social movements address disease, disability, or illness experience by challenging science on causation, diagnosis, treatment, and prevention in various settings. Scholars argue that embodied health movements in the United States are defined by three unique characteristics. First, they highlight the biological body in relation to social movements, for example, calling attention to the experience of embodiment in disability rights movements, women's health movements, and the environmental justice movement. Second, they argue that embodied health movements often include challenges to existing medical and scientific knowledge and to regulatory practice that are specifically tied to the human embodiment of health inequity based on difference of race, ethnicity, gender, class, and sexuality. Third, these movements involve activists collaborating with research and regulatory scientists and health professionals in pursuing treatment, prevention, research, and expanded funding. While the combination of these three characteristics makes embodied health movements unique, they are still similar to other social movements that depend on the development of a collective identity to mobilize action. That identity is often catalyzed when scientific institutions fail to offer disease accounts that are consistent with individuals' experiences of illness or when science and medicine offer accounts of disease that individuals contest.[57] Other scholars of social movements also focus on the personal and embodied experience of activists. As Sandra Morgen argues, "Too often the stories of social movements are told without enough attention to what the experience of being part of that movement meant to and felt like to those who participated in the movement. I don't believe we can understand the agency of political actors without recognizing that politics is lived, believed, felt, and acted all at once. Incorporating the experience of social movement involvement into analysis and theories about social movements may be difficult, but it adds a great deal to what we can learn about politics, social transformation, and political subjectivities."[58]

Environmental justice groups are taking a similar approach to climate and health; their embodied experiences motivate and inform their activism and are central to the politics of their movement. They are increasingly debating with experts over issues of truth and method in science. These groups are also demanding a greater role in environmental health decision-making that impacts their lives and bodies. Environmental justice groups are not only challenging the political use and control of science and expertise by claiming to speak credibly as experts in their own right; they are also challenging the process by which technical knowledge is produced. Conventional climate policy often overlooks the ways in which scientific

Table 2.1. Three spheres of science typically involved in movements for environmental justice

	Regulatory science	Research science	Street science
Truth claims	"Truths" to inform and legitimate public policy	"Truths" of originality and disciplinary significance	"Truths" relevant to the lived experiences of community members; policy advocacy; significance for action
Practicing institutions	Government; industry	Universities; government; industry	Community-based groups; social movement organizations; universities; government
Products	Studies and data analyses, often unpublished; private goods (with exchange value); public goods (with use value and value for democratic participation)	Published papers; public goods; private goods	Studies and data analyses; published and unpublished papers; press releases; community organizing, education, empowerment; public goods
Incentives	Compliance with legal requirements; balance interest and disinterest in economic development	Professional recognition and career advancement; legitimacy within the scientific field	Achievement of social justice and human rights; legitimizing the claims of social movement organizations and participating community members in the public sphere
Time frame	Statutory timetables; political pressure; funding and career constraints	Open-ended; funding and career constraints	Statutory timetables; political pressure; time demands of involved community members; funding and organizational constraints

Options in the face of uncertainty	Acceptance of evidence; rejection of evidence	Acceptance of evidence; rejection of evidence; waiting for more data	Acceptance of evidence; rejection of evidence; waiting for more data; precautionary/preventative; consensus over causes not necessary
Accountability	Congress or state legislatures; courts; media; targets of the findings	Professional peers; targets of the findings	Social movement peers; participating community institutions; courts; media; government; targets of the findings
Procedures for scientific review	Audits and site visits; judicial review; peer review (formal and informal), from "stakeholders" and "the public"	Peer review (formal and informal), from universities and research funders	Coproduction of knowledge; broadly defined "peer" review including nonscientists, social movement organizations, universities, regulators
Methodological and ethical standards	Absence of fraud or misrepresentation; conformity to approved protocol and agency guidelines; legal tests of sufficiency (e.g., substantial evidence, preponderance of the evidence)	Absence of fraud or misrepresentation; conformity to methods accepted by peer scientists; statistical significance	Absence of fraud or misrepresentation; alignment with the social movement principles and community expectations; legal tests of sufficiency; triangulation of statistical significance with lived experience of community members
Boundary actors, practices, norms, and objects	Each sphere of science has boundary actors (e.g., university and regulatory scientists) that can function in multiple spheres with credibility. As they move between boundaries, they bring along certain practices, norms, and objects that can influence a sphere of science.		

Source: Adapted from Liévanos et al., "Uneven Transformations," 204–5

knowledge and notions of expertise develop, become institutionalized, and tend to exclude from their cognitive domain other ways of knowing and doing.[59]

By extending the arena of legitimate climate change knowledge to include embodied knowledge, regulators and policymakers can better understand the insights that environmental justice advocates can offer to environmental problem-solving. In urban settings, where there is high uncertainty, embodied approaches can uncover new hypotheses rather than test predetermined ones. Embodied approaches, moreover, can provide a complex (or thick) description of the environmental condition that is faithful to the lived experience of residents. Such accounts provide a cultural consciousness that the environment can invoke multiple harms to human bodies and that the combining of knowledge and action for social change can ultimately help improve health in the most disadvantaged communities.[60]

However, embodied knowledge has its limits. It alone cannot ensure that policy is representative and democratic. Theorists of social movements have argued that activists are only successful in challenging environmental decision-making when they expend a significant amount of resources (financial, organizational, rhetorical, and cultural) to establish coalitions, develop knowledge frameworks, and act to change the terms of policy debates. For communities that lack capabilities and resources, the opportunities for embodied approaches to climate change are likely to be limited.[61]

Moreover, while embodied knowledge serves to empower disadvantaged communities, there are risks in overprivileging it. Anthropologists warn that romanticizing such approaches can result in knowledge that is not problematized. Not all local knowledge can be presumed to provide the best answers to environmental problems. John Briggs contends that the romanticization of embodied knowledge can result in "its adoption as the hegemonic knowledge system as a replacement for western science, making the same claims for pre-eminence."[62] Similarly, other scholars question whether there is a risk of "swinging from one untenable position ('we know best') to an equally untenable and damaging one ('they know best')."[63] The perspectives from the streets discussed in this book, however, are not intended to supplant scientific knowledge or the role of the scientific expert. Rather, they offer a framework for developing holistic, interactive, and multiparty processes of knowledge-making around climate policy. As Corburn has observed, a major challenge for the environmental justice movement is determining not whether to use the knowledge of scientific experts or knowledge from the streets "but rather how to recognize, based on the problem being addressed, new modes of collaborative inquiry that can capture the contributions from both local and professional science to improve environmental health for all."[64]

For nearly 40 years, environmental justice groups have challenged the impacts of the fossil fuel economy on low-income communities of color. Through the efforts of organizations such as Richmond's Asian Pacific Environmental Network (with offices in Oakland as well), they have shown that bodies can show signs of suffering and environmental inequality.[65] Their exposure risks are often most severe because multiple sources of pollution are more likely to aggregate and concentrate in such communities. Patterns of systemic inequity have motivated the environmental justice movement in Richmond and beyond to work across scales and sectors and to build knowledge networks and translocal coalitions. These emerging coalitions seek holistic and collaborative paradigms of environmental protection in which political, economic, cultural, and embodied contexts are fully considered. In the arena of climate change, their advocacy raises key questions: What impact will the environmental justice challenges to climate change policies deliver? Will their engagement expand our consciousness of the links between climate policy, pollution, health, and justice? Or will their tactics provoke opposition and deepen disagreements between actors seeking to solve climate change?[66] The next four chapters present detailed case studies of how an embodied, "from the streets" approach to climate change operates in practice, across scales and time. They highlight the ways in which environmental justice groups attempt to change the dominant environmental protection paradigm and improve climate-health decision-making.

CHAPTER THREE

Contentious Capitol Climates

IN APRIL 2006, I was working as a senior consultant for the youngest woman ever to serve in the California State Legislature. Elected in 2002, Cindy Montañez quickly became a rising star in the capitol. Over the years, she championed key legislative initiatives to clean up contaminated sites and improve air quality in the state's most disadvantaged communities. When nationally based environmental groups announced their intention to push for far-reaching climate change legislation, she took particular interest and asked me to meet with her. As I entered her office, she was seated at a large mahogany desk. Behind her were photos from her time at the University of California, Los Angeles, where in 1993 she led a 14-day hunger strike to protest a decision by the chancellor against the creation of a Chicano studies program. The hunger strike ultimately succeeded; the university relented and established the César E. Chávez Center for Chicana/o Studies. When I sat down, Montañez began to describe her frustration with traditional environmentalists whose global climate solutions often ignored environmental inequities in communities of color. Her comments went straight to the heart of the matter: "These environmental groups have a hard time understanding our communities because they haven't spent much time working there. So when they speak about environmental protection, they focus on global, faraway issues. But I don't see how a climate strategy that is disconnected from local communities at the frontlines of impacts is going to achieve anything real."[1]

This tension between global environmental protection and a local focus on the most disadvantaged communities was felt by other Latina/o lawmakers, whose growing political clout was transforming California's environmental policies. During this time, both demographic change and the impacts of climate change were reshaping California's political landscape. The frequency and severity of extreme weather events, heat waves, drought, wildfires, and polluted air were increasing, and low-income communities of color bore the brunt of these changes. Momentum from diverse constituencies was building, pushing legislators to commit to swift action. A statewide coalition of environmental justice groups pressuring legislators proclaimed, "We live in, work on behalf of . . . communities that are hardest hit by the failure to control climate change. . . . The opportunities in addressing climate change are great (green jobs, cleaner air, economic renewal, and a genuine transition to a clean energy future), while the risks of catastrophic climate collapse demand that we get the policy choices right."[2]

In the spring of 2006, as the Global Warming Solutions Act (AB 32) unfolded, Democratic Latina/o lawmakers joined environmental justice advocates to argue that while climate change is a global phenomenon, the state should develop climate solutions to benefit local communities. In these debates, global and local perspectives on climate change were embedded within differing economic models, systems of governance, and—in the broadest sense—worldviews. Disagreements over cap-and-trade became emblematic of these differences: traditional environmental groups, as well as the governor and state regulators, favored a market-driven solution focused on global greenhouse gas emissions. This perspective was shaped by neoliberal ideology that saw governmental regulation as a barrier to economic growth. Neoliberalism is a political and economic theory that has risen to prominence in market economies in the past half-century. Its proponents seek to reduce the role of the state in economic activity while granting the private sector more influence on both the making and execution of public policy.[3] The mainstream argument in favor of neoliberalism "is framed in terms of the efficiency of the market in contrast to the inefficiencies and high costs of government interventions."[4] Environmental justice activists, in contrast, supported direct government regulation and saw climate change as an issue linked with local air pollution and environmental degradation in neighborhoods. They advocated for bottom-up, transparent forms of decision-making and rejected the notion that market-based mechanisms could redress environmental inequities. Within an environmental justice worldview, cap-and-trade is a form of trickle-down environmental policy—it promotes geographically neutral emissions reductions and economic benefits that might eventually reach the most polluted communities.

Through a strategic alliance between Latina/o legislators and environmental justice advocates, a legislative compromise was reached that required California to explore various alternatives to cap-and-trade and ensure that climate action also yielded local health and economic benefits. The compromise provided a "justice of recognition," in which groups most affected by air pollution began to move from the margins of public discourse into a more prominent position of policy relevance. The resulting legislation, AB 32, went far beyond any other state in the country in reducing greenhouse gas emissions.[5]

When Republican governor Arnold Schwarzenegger signed AB 32 into law on September 27, 2006, traditional environmentalists allied with state leaders and agencies saw it as a turning point in international efforts to solve the world's most pressing environmental problem. The Natural Resources Defense Council and the Environmental Defense Fund shared the stage with the governor, the bill's authors, and a host of national and international dignitaries at signing ceremonies in Malibu and San Francisco's Treasure Island. Panoramic views of California's glistening coastlines served as symbolic backdrops, signaling the significance of environmental protection as recognized by the governor, state regulators, and traditional environmentalists. Yet environmental justice groups refused to join the ceremonies and were apprehensive about the implementation of the bill. Their skepticism was rooted in experience, which had led them to believe that state regulatory processes could be undemocratic. Polluters might well be allowed to determine where and how to reduce greenhouse gases and co-pollutant emissions without engaging the communities affected by those decisions.

When AB 32 moved to the implementation process at the California Air Resources Board, conflict arose over the essential questions of how to define climate change and integrate environmental justice into policy. In the hallways of the state's bureaucracy, climate change was defined through an exclusive focus on greenhouse gases and their effects on a global scale. This outlook, combined with a preference for market-based mechanisms, limited opportunities for creating public health benefits at the neighborhood scale. Tension between the law's purported focus on environmental justice and the perceived inequities of its cap-and-trade mechanism became a source of discord between environmental justice advocates, traditional environmental organizations, and state agencies. This conflict culminated in seven of the eleven state-appointed members of the board's Environmental Justice Advisory Committee joining a lawsuit against California that accused the state of failing to study alternatives to a cap-and-trade program. When a judge ordered the board to conduct a study, the board produced an analysis of alternatives while maintaining that cap-and-trade was the only feasible policy direction.[6]

In this chapter, I argue that the governor and state regulators implemented environmental justice concerns in a way that excluded the voices and embodied perspectives of communities of color. Through limited promises of economic development and health benefits in disadvantaged neighborhoods, state officials sought the approval of environmental justice communities. They used carefully managed public participation mechanisms, rather than open dialogue, to dissuade activists from opposing cap-and-trade. In turn, environmental justice groups perceived this behavior as manipulative and domineering. The result was a complete breakdown in relations between the board and environmental justice groups. At the heart of this conflict lay a politics of scale that privileged the global over the local and prevented integration of a "view from the streets" into California's climate change policies. This chapter illustrates the contentious nature of efforts to define climate change as a public concern and how the interaction of diverse kinds of knowledge and worldviews—as well as power and racial dynamics—shape contemporary climate governance and decision-making processes.

Latinos Legislating the Climate

> Climate change is a reality, and it puts in danger our sustainability. It should be of special importance to Latinos because Latinos for the most part live in the inner cities and are exposed to many carbon-based [and toxic] gases from industry, which not only impact the environment but public health.
>
> —California speaker FABIAN NÚÑEZ at the National Latino Congreso on September 11, 2006, shortly after the state legislature passed AB 32

The person behind the state's landmark law to curb climate change does not fit the stereotypical image of a California environmentalist. Assembly Speaker Fabian Núñez is a former competitive boxer and labor organizer with a view of environmentalism that, through embodied experience, links climate change to a public health context. As a child growing up in the low-income Latina/o immigrant communities of San Diego, Núñez saw firsthand the impact that residents felt from poor air quality. Children in these neighborhoods suffer from disproportionately high rates of asthma and respiratory diseases. Núñez's rise to power as leader of the state assembly disrupted the notion of California environmentalism as a luxury of the elite. As described by senior capitol staff members and traditional environmentalists, Núñez placed pressure on the broader environmental movement to look beyond white, affluent supporters and make themselves relevant to legislators and communities of color if the movement was to survive.[7]

A turning point occurred in April 2006, when Assembly Member Fran Pavley, a respected environmentalist from a wealthy, predominately white coastal district that included Malibu, approached Núñez for help on her bill, AB 32. For nearly two years, the measure had languished amid strong opposition from fiscally moderate Democrats. Chief among these were members of the California Latino Legislative Caucus. California's new demographic reality as a majority-minority state made the caucus influential in passing important environmental legislation. It had grown beyond its traditional urban population base, and its newest members hailed from politically moderate rural and suburban districts in the San Joaquin Valley and Orange County.[8] Many of these members were from competitive districts where Democratic voter registration led that of Republicans by only a few percentage points. These legislators had to engage in a delicate balance between environmental stewardship and promoting economic development to remain in office. Latina/os representing these communities often responded to industry arguments about the cost of environmental regulation—either in the form of increased energy costs for residents or the loss of blue-collar jobs in their region.[9]

As the legislature was changing, Pavley had difficulty building close relationships with the new lawmakers. Caucus members described her as reserved and unable to relate to individuals who didn't share her approach to environmental protection. This failure to connect was problematic at a time when Latina/o legislators were seeking unconventional methods to link global policy to a local context. Another challenge was that Pavley's legislation was sponsored by two national and traditional environmental organizations, the Natural Resources Defense Council and the Environmental Defense Fund. Members of the caucus regarded both as "elitist" institutions whose policies did not address the consequences of environmental regulations for low-income workers and residents.[10] Latina/o legislators from competitive districts were reluctant to risk supporting controversial legislation that was unlikely to improve environmental and economic conditions in communities of color. An environmental justice advocate noted this disconnect between Latina/o legislators and mostly white environmentalists as a barrier to advancing progressive legislation: "The number of districts that are being represented by people of color has doubled, tripled in the last ten years. But in terms of support for the environment, a lot of the representatives are . . . on the fence and susceptible to being influenced by the oil industry or Chamber of Commerce . . . because they haven't heard of what are the environmental benefits for their districts. . . . Traditional environmentalists don't have any history in these districts; they are not relevant; they are greeted with skepticism. The California Chamber of Commerce is very persuasive because they frame [environmental] programs as job killers."[11]

Understanding these political dynamics, Núñez offered to shepherd AB 32 through the legislature. However, he had one key condition

—that he would take over as its lead author. At that point, both capitol staff members and traditional environmentalists noted, AB 32 was "essentially dead" and could not proceed without Núñez's political leadership.[12] He was seen as a skilled tactician who could engage with a Republican governor and moderate Latina/o members to get the votes needed for passage. A Latino Caucus member reasoned, "That type of negotiation was something environmental organizations, with Pavley as their legislative champion, were not able to do. They weren't able to negotiate elements that could protect community needs, jobs, the environment, and bring diverse stakeholders to the table."[13] Without other options available to her, Pavley reluctantly agreed to cede the main authorship of AB 32.

Mindful of existing political realities, Núñez sought to recraft the legislation to address the diverse needs of members, the Latino Caucus in particular. He knew that focusing AB 32 solely on greenhouse gas emissions and global environmental protection was not enough. During an election year, such an approach could place members in competitive races at risk of voting for a bill that powerful business interests condemned as a "job killer." The California Chamber of Commerce releases an annual list of "job killer" bills to identify legislation that it believes will "decimate economic and job growth in California." The chamber tracks the bills throughout the legislative session and works to "educate legislators about the serious consequences these bills will have on the state." In 2006, AB 32 was the chamber's top target for defeat.[14] With these constraints in mind, Núñez worked with key environmental justice advocates and the Latino Caucus to help rescale the debate of climate change to focus on two elements: public health, particularly air pollution and respiratory diseases that California's low-income communities of color face at epidemic levels; and local job creation and economic opportunity. These efforts targeted moderate Latina/o members who were considered "business Democrats" and had deep concerns about the impact on jobs as the state sought to transition from polluting industries to a more sustainable economy.[15]

Working with legislative allies was a key strategy for environmental justice activists. At the time, California environmental justice groups were like others around the country, in that they were locally based with limited resources. They lacked full-time lobbyists in Sacramento who could work on issues or advocate with policymakers; as a result, they were often disadvantaged in influencing comprehensive policy reforms. Latina/o legislators described early environmental justice campaigns as sporadic, with weak political influence. Nevertheless, by 2006, more than 20 incremental state laws addressing a range of environmental justice issues had been enacted. Environmental justice groups used two approaches to advocate at the state level. The first was to build partnerships through the California Alliance Working Group on Environmental Justice, which comprised locally based organizations: the Asian Pacific Environmental Network, the Center for Community Action and Environmental Justice, Communities for a Better Environment, Environmental Health Coalition, and People Organizing to Demand Environmental and Economic Rights. The second was a collaboration with expert organizations that, while not rooted in local communities, conducted policy analysis and advocacy on behalf of the environmental justice movement. These organizations included California Communities against Toxics; the California Rural Legal Assistance Foundation; the Center on Race, Poverty, and the Environment; and the Environmental Rights Alliance.[16]

Such approaches were useful for incremental campaigns but not the development of comprehensive climate change policy legislation. AB 32 was a fast-moving proposal, and policy responses were geared toward global objectives, not local action. The complex nature of climate change and pressure for immediate policy action, moreover, intensified power differentials between environmental justice groups and well-resourced entities, such as the Natural Resources Defense Council and industry groups. As pointed out by Angela Johnson Meszaros, an environmental justice advocate with Earthjustice, "One of the issues that makes this climate work really different than some of the other issues I've worked on, is this frame of immediacy. It's really dynamic, fast-moving, high-paced policymaking which makes it another level of challenge for organizations that are small, underfunded, and otherwise under-resourced, without access to decision-makers and information that is useful and relevant."[17]

To help compensate for these challenges, the Environmental Defense Fund hired Rafael Aguilera, a Latino lobbyist who had worked in the state capitol for various elected officials. Aguilera targeted his lobbying efforts to moderate Latina/o lawmakers by stressing the links among localized climate change, job creation, and public health. The Environmental Defense Fund described Latina/o members as often supportive of industry interests and skeptical of the benefits of environmental policies. According to their newsletter, "Advocates for minority communities don't always see eye to eye with environmentalists. So, it wouldn't have taken much to derail the Global Warming Solutions Act, especially during an election year."[18] According to a representative with a traditional environmental group, Aguilera was motivated by the realization that environmental justice issues were key to persuading moderate Latina/o legislators that AB 32 would benefit their constituents. His involvement was also needed to ease tensions created when Núñez floated draft bill language mandating cap-and-trade. The amendment was requested by Governor Schwarzenegger, who threatened to veto legislation not containing it.[19]

Cap-and-trade is a market-based regulatory program intended to limit the total amount of greenhouse gas pollution from regulated entities (fig. 3.1). The program creates a previously nonexistent market price for pollution. According to the Environmental Defense Fund, "The cap on greenhouse gas emissions is a limit backed by science. Companies pay penalties if they exceed the cap, which gets stricter over time. The trade part is a market for companies to buy and sell allowances that permit them to emit only a certain amount. Trading gives companies a strong incentive to save money by cutting emissions."[20]

Environmental justice groups such as the Environmental Rights Alliance, the Asian Pacific Environmental Network, and California Communities against Toxics opposed cap-and-trade and threatened to voice their concerns to Latina/o members who were on the fence over supporting the controversial legislation. Environmental justice groups argued that a cap-and-trade system could create geographically uneven reductions in greenhouse gas emissions and co-pollutants, thus limiting the opportunity for health benefits at the neighborhood scale. Companies with high-polluting facilities, they claimed, could buy pollution allowances from other companies that had not exceeded their caps and continue emitting both carbon dioxide and co-pollutants that degrade local air quality. Firms with the most an-

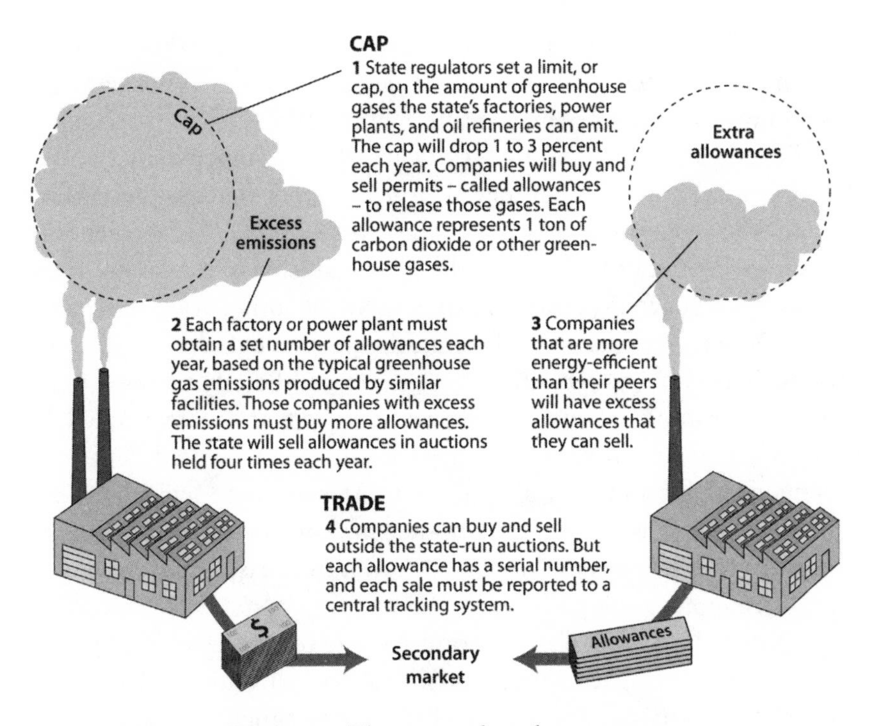

Figure 3.1. The cap-and-trade process.
Todd Trumbull/*San Francisco Chronicle*

tiquated facilities are generally seen as emitting the most emissions and are often located in low-income communities of color. Environmental justice advocates asserted that these companies were likely to purchase the most emissions allowances.[21] Jane Williams, with California Communities against Toxics, argued that instead of motivating change through a consistent and transparent price signal and direct emissions reductions at the source, reliance on pollution trading would reward polluters but provide little public benefit.[22] Cap-and-trade, furthermore, was a central concern for environmental justice groups because the industrial sector's (including cement factories, refineries, and oil, gas, and electricity production facilities) obligations to greenhouse gas emissions reductions was proposed to come primarily through its compliance under the program.[23]

To support their claims, environmental justice advocates cited two studies of the Regional Clean Air Incentives Market, an emissions trading program developed to reduce nitrogen oxides emissions in Southern California. The first study indicated that this program actually increased nitrogen oxides emissions in Wilmington, a working-class, Latino immigrant neighborhood of Los Angeles, while reducing emissions on a regional scale. The other study showed that four oil refineries purchased large numbers of emissions credits under the program instead of installing pollution reduction technologies in Wilmington. As a result, local residents and workers were exposed to benzene (a human carcinogen) and other volatile organic compounds that could have been mitigated through the use of readily available pollution reduction technologies. Based on these examples, environmental justice advocates argued, the state needed to directly regulate emissions sources in ways that would address both climate change and toxic air pollution. An exclusive focus on greenhouse gases through cap-and-trade, they stated, could worsen air quality in environmental justice communities despite improving it at the regional scale.[24]

Núñez took such claims seriously. The registered opposition of environmental justice groups provided wavering fiscally moderate legislators with a convenient excuse for not supporting AB 32. To appease environmental justice groups and prevent the loss of key votes, Núñez rejected the draft provision mandating a cap-and-trade system, despite the governor's veto threat. The Democratic legislature and Republican governor eventually negotiated a compromise that instead included a statewide emissions limit for 2020 and delegated to the California Air Resources Board key decisions over implementing the state's greenhouse gas emissions reduction strategies. This gave the board discretion to adopt market-based mechanisms or direct emissions reductions at polluting sites (also known as "command-and-control" regulations), which environmental justice groups preferred. The latter would rely on government regulation (permission, prohibition, standard-setting, and enforcement) as opposed to financial incentives.[25]

To further address public health and environmental justice concerns, Núñez included additional provisions in the bill. These required the board, to the extent feasible, to "consider" cumulative emissions impacts in affected communities and design any market-based mechanism to maximize community benefits and prevent increases in local emissions of toxic air pollutants. With the princi-

Table 3.1. Environmental justice and AB 32

Environmental justice concepts	Corresponding legislative language in AB 32
Pollutant reduction	Authorizes but does not mandate the board to adopt market-based mechanisms to comply with AB 32 regulations Requires the state to consider various alternatives in addition to a cap-and-trade system
Public health connection	Prior to the inclusion of any market-based mechanism, the board must do the following: 1. Consider the potential for direct, indirect, and cumulative emissions impacts from these mechanisms, including localized impacts in communities that are already adversely impacted by air pollution 2. Design any market-based compliance mechanism to prevent any increase in the emissions of toxic air contaminants or criteria air pollutants 3. Maximize additional environmental, health, and economic benefits for California, as appropriate
Public participation	Requires the institutionalization of the Environmental Justice Advisory Committee to advise the board on the implementation of AB 32 Requires public workshops to be held in regions that have the most significant exposure to air pollutants
Community empowerment	Requires opportunities for disadvantaged communities to participate in and benefit from public and private investment from the greenhouse gas reduction programs established by AB 32 regulations

Source: California Health and Safety Code § 38591

ple of equity driving their efforts, environmental justice advocates secured two more provisions in AB 32 (table 3.1). The first required the formation of the Environmental Justice Advisory Committee to advise the board on implementation of the "scoping plan," a proposed framework for achieving greenhouse gas reduction targets. The inclusion of such a committee was not unique to environmental governance. But its codification as one of two committees mandated in the bill was symbolically significant (the other was the Economic and Technology Advancement Advisory Committee), as was the composition of its membership. The committee included representatives from California communities with significant exposure to air pollution, with particular attention paid to minority and low-income communities. The second provision, known as the Community Empowerment Amendment, was intended to allow disadvantaged communities to participate in and benefit from the AB 32 greenhouse gas reduction plan. It mandated the board to direct public and private investment, such as green jobs and renewable energy projects, to environmental justice communities.[26]

The inclusion of these elements represented a watershed moment for Latina/o legislators, as well as for the environmental justice movement. AB 32 was the first time that Latina/os had played a significant role in crafting comprehensive environmental policy. Until this point, white legislators had dominated the policy arena of natural resources protection. A Latino Caucus member explained, "We weren't doing other issues because traditional environmental groups wouldn't work with us. They would always go to Fran Pavley for everything. Núñez just took over, and Pavley wasn't at the center of the discussion anymore."[27] According to a member of a traditional environmental organization, this displacement caused tension with some colleagues who felt left out by the new political order. They thought that Núñez had taken over the bill to connect unrelated issues, such as poverty alleviation and local pollution, to the problem of global climate change.[28] Despite these tensions, another Latino Caucus member argued, it "was a necessary political maneuver [to take over authorship]. They considered us too business friendly. If we did not have a 100 percent voting record with them, traditional environmental groups would not engage us. What was different about Núñez, he was willing to engage everyone." The environmental protection paradigm at the time was perceived as not aligning with Latina/o legislators' values and perspectives, and it took a strategic approach to shift the dynamics.[29]

Carbon Markets and Neoliberal Latinos

Given the politics and socioeconomic conditions of their districts, members of the Latino Legislative Caucus were concerned about questions of public health and pollution, as well as greater economic development. Despite individual differences, many members were open to the neoliberal arguments that traditional environmental organizations adopted to support AB 32. They focused on markets, innovation, and economic growth to refute the views of groups such as the California Chamber of Commerce and the California Farm Bureau Federation, which opposed AB 32. Within this neoliberal framework, the cap-and-trade system, which environmental justice activists saw as flawed and inequitable, became central to supporters' advocacy for the bill.

To torpedo support for AB 32, business groups began a multimillion-dollar campaign that included radio advertising and targeted mail in competitive legislative districts. They argued that AB 32 would force companies to leave California to avoid onerous regulations or compel businesses to curtail operations and displace workers to meet them. The business lobby further warned that the bill would increase energy prices as utilities sought cleaner energy sources. To refute these claims and insulate vulnerable Democratic caucus members, Núñez mounted a counter campaign with the aid of the Environmental Defense Fund and the Natural Resources Defense Council. They sought to convince cautious legislators that AB 32 would spur innovation and uplift California's economy, reprising the state's early-adopter success in the biomedical and high-technology industries.[30]

Núñez and his allies enlisted the support of influential venture capitalists and Silicon Valley entrepreneurs. They argued that new climate change regulations could create a boom in industries such as solar power and biofuels that would propel the state's economy for decades. To support these claims, AB 32 proponents circulated a University of California, Berkeley, economic study that estimated that greenhouse gas emissions reductions could create 17,000 new jobs and add $60 billion to the gross state product by 2020. Governor Schwarzenegger supplemented these efforts by creating a special task force, the Climate Action Team, which was charged with identifying cost-effective methods to reduce greenhouse gas emissions. This task force found that a cap-and-trade program could add more than 80,000 new jobs over the next several decades.

Responding in part to the promise of jobs in their local communities, 27 out of 28 members of the Latino Caucus voted for the climate legislation, and 21 signed on as coauthors.[31] The campaign to link AB 32 to the green economy was so successful that it eclipsed the public health benefits the measure could bring to disadvantaged communities. During the final vote for AB 32 on August 30, 2006, the majority of assembly members speaking in favor of the legislation commented on its potential for protecting the environment while creating thousands of new jobs and billions in investments within California.[32]

AB 32 passed the assembly almost on a strict party-line vote. Its linkage to the green economy, however, set the stage for the bill's implementation on neocommunitarian terms. Neocommunitarian strategy uses limited forms of community engagement and the promise of economic development to gain support in disadvantaged communities. While activist discourse often cites environmental justice values as a challenge to market-driven environmental policy, neocommunitarian approaches seek to make those values compatible with neoliberalism.[33]

Ryan Holifield's case study of the US Environmental Protection Agency's (EPA) Superfund hazardous waste site remediation program provides an example of the inherent contradictions that a neocommunitarian strategy can bring to environmental justice communities. The EPA developed job-training programs for cleaning up hazardous waste and provided businesses with incentives to redevelop contaminated sites. However, as implemented, the agency's neocommunitarian strategy also prevented open and substantive engagement from the streets. Decisions were made from the top down, and local residents had limited opportunities to shape policy and programs. According to Holifield, the EPA chose not to adopt environmental justice principles of community empowerment or community-based decision-making. Instead, it instituted public participation as a form of "improved public relations," making the agency's decisions more accessible and allowing the residents to submit recommendations through carefully managed community involvement. Many environmental justice groups throughout the country perceived the remediation program as an attempt to co-opt environmental justice values in the service of a neoliberal agenda.[34]

The neocommunitarian strategy in the context of global climate change can be even more problematic for environmental justice communities. The contested politics of scale heightens power differentials, as environmental justice groups are unable to translate their local values to regulators focused on global policy and benefits. Nonetheless, the California legislature set the stage to follow the EPA's lead in framing market-based mechanisms as compatible with environmental justice goals. Núñez made this commitment clear on August 30, 2006, in his closing remarks before the passage of AB 32 and in a speech one week later at the National Latino Congreso (a major summit of Latina/o elected officials and community leaders). On the assembly floor, Núñez asserted the virtues of neoliberal approaches and spoke as if cap-and-trade were a foregone conclusion. The bill, he said, had the support of "good corporate citizens" with existing investments in green technologies "that will ultimately put more equity in their pocket when this market mechanism kicks in and we develop a cap-and-trade system."[35]

In his address to the National Latino Congreso, however, Núñez made no mention of the cap-and-trade program. Instead, he linked AB 32 to community economic development and the health of residents affected by air pollution. Arguing along neocommunitarian lines, he framed AB 32 as an opportunity to promote job creation in Latino communities: "I'm excited to tell you leaders from all over the country what we're doing here in California to slow climate change. Because I know our success means you can do it too. And that not only means better health for our planet, it also means more jobs and opportunities in our communities as we develop the green technologies to combat climate change."[36] This targeted combination of market-based and neocommunitarian arguments helped unite support for AB 32 and ensured its enactment. But the dominance of those arguments also contributed to later conflict over implementing the bill's provisions. The neocommunitarian approach to public consultation antagonized environmental justice groups because they saw it as marginalizing their concerns. When AB 32 moved from the legislature to the regulatory implementation process at the California Air Resources Board, it became clear that Governor Schwarzenegger had already decided that cap-and-trade was the best route to reducing California's greenhouse gas emissions. And it seemed he was willing to advance it by any means necessary.

Conflict over Implementation of AB 32

Throughout the AB 32 negotiations, Schwarzenegger supported the economic benefits of a market-based system without regard to its possible impact on environmental justice communities and even threatened to veto any legislation that did not mandate cap-and-trade. Eager to claim credit for global climate leadership, the legislature and the governor took an expedient approach and enacted AB 32 without settling their differences over how to implement the new law. The legislation instead delegated key implementation decisions to state regulators.[37]

When AB 32 reached the implementation stage at the board—an agency for which the governor has oversight authority—Schwarzenegger initiated efforts that directly undermined the bill's environmental justice provisions. On October 16, 2006, less than a month after signing AB 32 into law, the governor issued Executive Order S-20-06, which declared cap-and-trade the most cost-effective mechanism to reduce California's greenhouse gas emissions. The executive order instructed the board to develop a comprehensive, market-based compliance program, while the California Environmental Protection Agency was directed to establish the Market Advisory Committee to advise the board on the program's formulation.[38]

Environmental justice advocates took Schwarzenegger's executive order as a betrayal. To them, it confirmed that the inclusion of environmental justice elements in AB 32 was mere "tokenism" to prevent them from opposing the measure. According to a senior Schwarzenegger appointee, when the board convened the Environmental Justice Advisory Committee to advise on implementation of the scoping plan, a lack of trust was evident, and conflict ensued over cap-and-trade. Activists further characterized Schwarzenegger's actions as creating a regulatory atmosphere incompatible with environmental justice goals.[39] In June 2007, the governor fired the board chair, Robert Sawyer, a respected scientist and academic, for attempting to implement greenhouse gas reduction strategies before the adoption of a cap-and-trade system. The measures that Sawyer sought included dozens of ideas favored by environmental justice groups, such as higher fuel efficiency standards for heavy-duty vehicles and diesel construction equipment. After Sawyer's firing, the board's executive officer, Catherine Witherspoon, resigned in protest. She accused the governor of pushing a cap-and-trade strategy "to find ways to reduce costs and satisfy lobbyists," although other regulatory methods were needed to meet the law's goals.[40] Angela Johnson Meszaros, cochair of the board's Environmental Justice Advisory Committee, characterized Sawyer's dismissal as a major loss for the environmental justice movement.[41]

Núñez criticized both the executive order and Sawyer's termination, which he called abuses of power that weakened key environmental justice provisions. The legislature, in response, held oversight hearings to determine if the governor was improperly influencing the board. They found an unprecedented level of interference in board business yet no evidence of legal wrongdoing. Throughout the controversy, Schwarzenegger maintained that "the only way to fight global warming is through a cap-and-trade system." In speaking of Sawyer's replacement, he argued, "Whoever gets the position has to believe in that. He has to be tough, even though they'll hear people whining. I've heard people whining. But, we've got to be extremely sensitive toward businesses here."[42]

By July 2007, perhaps to quiet criticism that he was obstructing AB 32 implementation, Schwarzenegger selected the respected environmental lawyer Mary Nichols to lead the board. As a former appointee of two Democratic governors and President Clinton, Nichols received approval from both business and traditional environmental groups for her pragmatic approach to air quality governance. Environmental justice groups, however, were skeptical. Nichols had played a key role at the US EPA in promoting market-based mechanisms to control pollution. On the same day that the governor appointed Nichols, he also selected a former oil industry and utilities lobbyist, Cindy Tuck, as undersecretary of the state's Environmental Protection Agency. In this capacity, Tuck would help direct the activities of the Market Advisory Committee in its recommendations for establishing a cap-and-trade system.[43]

Schwarzenegger's alignment with industry and preference for cap-and-trade foreshadowed conflict throughout the implementation of AB 32. Between July 2007 and April 2009, during more than a dozen formal committee meetings, the board and the Environmental Justice Advisory Committee fought over the development of the scoping plan and its effect on low-income communities of color. Committee members viewed the board's actions as employing public participation as a means to pacify aggrieved communities without addressing how the implementation of AB 32 might alleviate or exacerbate environmental injustice at a structural level. Compared to the Market Advisory Committee and other technical groups advising on AB 32 implementation, the Environmental Justice Advisory Committee received limited staff support and resources. This lack of support inhibited the committee's efforts to turn AB 32's Community Empowerment Amendment from a goal into concrete policy actions.[44] Jane Williams, the cochair of the committee, described the board's process as dismissive: "The tone and tenor of those meetings has adopted an eerily similar format. We express the grave concerns of the committee, make suggestions on how the agency can address the concerns, staff listens intently and indicates that they will consider our concerns and get back to us. Time passes, we express the concerns again; staff listens intently, and say they will address our issues in the future."[45]

The AB 32 public workshops, especially those held in low-income communities of color, were purported to be conducted with limited board resources. While the committee issued advisory reports to the board, many of its deliberations were not documented. Its meetings and the public workshops were not translated into Spanish or other languages commonly spoken in environmental justice communities. One committee member described the board's public participation process: "There's no transcript, . . . there's no real record of what we talked about . . . which is somewhat disturbing."[46]

Facing a lack of state resources, the Environmental Justice Advisory Committee supplemented its work with funding from private foundations. This support enabled members to produce hundreds of pages of policy recommendations for the scoping plan. The committee promoted traditional command-and-control mechanisms that relied on direct government regulation and incentives. A key recommendation was a tax per ton of carbon dioxide, which would be assessed at a low rate initially and would rise over time. The committee viewed a carbon tax as producing the most geographically equitable carbon reductions by creating immediate economic incentives for polluters to invest in cleaner technologies. A carbon tax, the committee claimed, would also be more transparent and harder to evade. The revenue would be subject to public disclosure, making it clear which entities were complying with the regulation and how. In contrast, under cap-and-trade, the specific entities trading and selling pollution permits would not be subject to the state's public disclosure laws; such information is considered to be confidential and trade secret.[47]

The Environmental Justice Advisory Committee's main objection to cap-and-trade was that it excluded community voices from important regulatory decisions. Communities would be prohibited from the trading of allowances that could shift the spatial distribution of reductions in greenhouse emissions and co-pollutants. According to Sean Hecht, under command-and-control regulations, "local communities can influence the regulatory process by participating in permitting proceedings . . . or intervening in direct administrative or court enforcement actions."[48] By contrast, under cap-and-trade, polluters, investors, and traders form the system that determines the regulatory outcome on a local level by controlling available carbon emissions allowances. This sense of exclusion and lack of transparency was further exacerbated by the board's decision to incorporate Western Climate Initiative Inc., the company that would manage the cap-and-trade auctions, 3,000 miles away in the state of Delaware. As an entity outside California, it would be exempt from California's open government meetings and public disclosure laws.[49]

These concerns led members of the committee to join a statewide environmental justice coalition in a declaration against cap-and-trade. The declaration stated that they would fight at every turn against a carbon-trading system that would exacerbate pollution in disadvantaged communities: "Carbon trading is undemocratic because it allows entrenched polluters, market designers, and commodity traders to determine whether and where to reduce greenhouse gases and co-pollutant emissions without allowing impacted communities or governments to participate in those decisions."[50]

Environmental justice advocates held that relying on the market to reduce pollution was impractical and would reinforce existing racial inequalities. In their recommendations for the scoping plan, the Environmental Justice Advisory Committee expressed an explicit connection between race, place, and carbon markets. Committee cochairs stated in a letter to the board, "It is market-based decisions, within a framework of structural racism in planning and zoning decisions, which has created the disparate impact of pollution that exists today; relying on that same mechanism as the 'solution' will only deepen the disparate impact."[51]

Scholars of environmental justice have supported this perspective and argue that since the early part of the last century, zoning and planning have been used for more than just policing land use. They are also methods of regulating where people live and segregating classes of people. In effect, "environmental racism is found at the intersection of these two uses." The economist Yale Rabin studied the racial history of zoning throughout the United States and found that predominately white decision-making bodies systematically "downzoned" the classification of established residential communities of color to site industrial uses.[52] As a result, African American neighborhoods were often placed within industrial areas or near noxious land uses: "While de facto segregation is no longer legal, the land use patterns (and the industrial zoning) created by these decision-makers still controls many communities across the country."[53] Environmental justice advocates posit that in the context of climate policy, cap-and-trade will perpetuate these conditions—factories will expand their operations near communities of color or will be more likely to locate new facilities near them, rather than near white neighborhoods. Communities of color would bear a continued or aggravated burden of exposure to co-pollutants.[54]

The Politics of Scale and Justice

The implementation of AB 32 manifested profound tensions among the politics of race, carbon markets, and scale. Throughout the scoping plan process, the board sought to limit the range of measures and to address only global greenhouse gas emissions. Regulators argued that the "cap" part of the cap-and-trade system was inherently equal. Regardless of location, everyone could benefit from the cap on greenhouse gas emissions because their effect is global. In contrast, the Environmental Justice Advisory Committee held that the "trade" part was not equal. Greenhouse gas reductions and related co-pollutant reductions could have uneven geographical distribution. While statewide emissions might decline, the burden for overly polluted communities of color could potentially worsen or improve at rates lower than statewide. As a result, the promised localized health benefits for disadvantaged communities may not materialize (they argued this was a form of trickle-down environmental policy).[55]

Board officials rejected the Environmental Justice Advisory Committee's local scale of climate action and dismissed it as grounded more in emotion than in science. The board's technical advisory committees perceived the use of only command-and-control approaches as stifling greenhouse gas reductions targets and economic development. They argued that a carbon tax would be cost-prohibitive to administer while limiting opportunities for innovation in the green economy through inflexible and complicated rules.[56] Dan Skopec, an assistant secretary at the state's Environmental Protection Agency, said that the committee's climate strategy was self-defeating: "Using the umbrella of global warming to satisfy other agendas is really going to distract from the solution and create inefficiency. So as we go forward, I hope that we can all focus this effort on the problem of reducing greenhouse gases and not try to solve everyone else's unsolvable problems in other areas."[57]

The Natural Resources Defense Council and other traditional environmental groups, invoking neocommunitarian arguments, claimed that a well-designed cap-and-trade system could meet AB 32's global greenhouse gas reduction goals and address local environmental justice concerns. They produced an analysis showing that air pollution and health risks would decrease on a statewide scale, including in communities of color, under the board's scoping plan. The council estimated that the plan could prevent more than 700 premature deaths and thousands of other negative health impacts annually, saving up to $5 billion in health costs by the year 2020. The board produced similar public health analyses in an attempt to address the Environmental Justice Advisory Committee's arguments against cap-and-trade.[58]

Members of the Environmental Justice Advisory Committee dismissed these studies, which they said lacked both a rigorous methodology and a scientific peer review. They argued that co-pollutant emissions reductions would largely occur at the regional scale, with a small portion of health benefits trickling down to environmental justice communities. Yet the committee could not produce data to validate its claims. While many universities and nonprofits have partnered with environmental justice groups in the last 20 years to develop health and scientific analyses to influence public policy, such partnerships were not part of the scoping plan process. The committee's lack of evidence, in turn, bolstered the board's legitimacy in decision-making. The board took on the role of technically disciplined expert and depicted the committee as biased advocates offering unsubstantiated positions.[59]

In the policymaking process, credibility is essential in the validation of knowledge. To gain credibility within civic decision-making structures, environmental justice advocates are expected to translate their embodied experience and produce rigorous scientific data to prove claims. However, according to the sociologists Daniel Faber and Deborah McCarthy, the limited financial and technical resources of environmental justice groups—arguably one of the most underfunded social movements in the United States—makes such expectations difficult to achieve. The vast majority of environmental justice groups have fewer than five paid staff members each; many still function as volunteer organizations. Unlike larger environmental groups such as the Sierra Club, environmental justice groups lack a dues-paying membership to offset the cost of technical experts who can aid advocacy campaigns.[60] In the context of AB 32's implementation, this asymmetry was further exacerbated by the relative lack of state support. In a letter to the board chair, Mary Nichols, the committee objected that there was "no request for proposals or other efforts to seek outside expertise on how best to understand, research, and answer" their range of questions. As a result, the committee argued, the board could not evaluate the total costs and public health impacts of the scoping plan on disadvantaged communities.[61]

This perspective on the scoping plan, as a technical process driven by power relations, was supported by a former Latino Caucus member who became an industry lobbyist: "The plan was created by industry experts, scientists, researchers, board staff, and the governor's office. There was a lot of money on the lobbying side. Lobbying firms were hired to represent the groups that had power."

As a lobbyist, the former caucus member retained attorneys and experts to influence the plan: "The minute I walked into any meeting with the board, it was so complicated and technical. . . . Unless you had the resources of a multimillion-dollar business, you could not engage effectively."[62]

For environmental justice groups, the implementation of AB 32 demonstrated how climate change policy is embedded within dynamics of power. Scale, in particular, represented a power struggle among different actors to reframe a policy issue to their own advantage. In this process, they "create hierarchies among levels—that is, assert the primacy of a particular level" over another. Defining the scale of a problem involves decisions about which aspects are significant, and which are not.[63] Thus scale can be used by governing institutions, such as the state, to control disenfranchised groups by regulating them and their environmental problems to a manageable measure.[64] In a process identified by James C. Scott as "state simplification," governments appeal to larger interests as they simplify diverse local systems and then use the newly unified systems to rationalize development planning and environmental management. People and nature are addressed at a particular scale and level of resolution at which phenomena are deemed understandable. In this process, the state's capacity to dictate the policy and legal representation of scale often exceeds the capacity of individuals at the neighborhood level.[65]

This power differential is further supported by the state control of data, analytical methods, and dissemination of information. The capacity to control and capture resources from different scales underpinned the board's policymaking process. The implementation of AB 32, said a senior board official, was intentionally focused on global greenhouse gas reduction targets and not local regulation: "We have a whole array of programs that deal with a variety of different kinds of contaminants . . . and we don't want people to somehow get into a mind-set where they think that AB 32 is the tool they have to use to deal with those."[66] Exacting a level of state control over pollution and risk issues, however, is often at odds with the "scales of everyday experience" that California environmental justice advocates encounter. Their embodied knowledge is derived from their continual exposure to pollutants and from other socioeconomic inequities in their local communities. By challenging the legitimacy of climate change policy that operates at a global scale, the geographer Hilda Kurtz contends, environmental justice groups are exposing "the ways in which the social construction of scale shapes and is shaped by political and economic processes." In this way, scale becomes a contested tool for understanding and representing climate change.[67] Although climate change presents particular challenges to environmental justice advocacy, the need to engage with scale as a relation of power is not new. Other scholars argue that the environmental justice movement is situated between local scales at which the community protests unwanted pollution and the broader geographic scales where pollution can be resolved. Appeals for environmental justice, therefore, are a strategic response to the opportunities and constraints of regulatory procedures that exist within a particular spatial and political context.[68]

The California Air Resources Board operated within these established systems of knowledge-making and governance as it developed the scoping plan for AB 32. On December 12, 2008, bolstered by what regulators deemed scientifically and economically sound evidence, the board adopted a plan focused on reducing greenhouse gas emissions at the global scale. It argued that cap-and-trade was the mechanism best aligned with the legislative intent of AB 32. Rather than mandating a carbon tax, the board contended that the flexibility of cap-and-trade would help identify "where and how" emissions reductions could be achieved most cost-effectively. In turn, this flexibility could stimulate the development of new green technologies and jobs.[69] The board's confidence in the power of markets to create positive change represents part of a continued retrenchment of direct government regulation. California has become an important site of contestation over neoliberal approaches to environmental governance. In conflicts over the administration of climate change, opposing positions are often oriented through divergent worldviews of climate action (table 3.2).[70] The first worldview centers on state regulators' goal of greenhouse gas reductions at the global scale; the second, on the localized emissions reductions that environmental justice groups seek.

Facing insurmountable differences between worldviews and unable to negotiate the boundaries of the board's global perspective, 7

Table 3.2. Divergent worldviews of climate action

	California Air Resources Board	Environmental Justice Advisory Committee
Carbon pricing mechanism	Markets establish value of carbon allowances via a cap-and-trade system.	State intervenes directly to set carbon tax rate.
Means of reducing emissions	Market choice provides flexibility for polluters and incentives for greenhouse gas emissions reductions.	Command-and-control regulatory methods impose specific, inflexible emissions limitations.
Geographic scale of action	Greenhouse gas emissions reductions calculated on a statewide basis contribute to global total.	Greenhouse gas emissions should be reduced on-site, with emphasis on polluters in disadvantaged communities.
Reduction of co-pollutants	Only greenhouse gas emissions reductions are required; reduction of co-pollutants will be "considered" where feasible.	Greenhouse gas emissions reductions measures should coincide with the reduction of co-pollutants.
Regulatory science process	Regulatory science is established by the board and an expert peer-review process.	Embodied knowledge is fundamentally important. Participatory research methods are favored.
Co-benefits	Market efficiency, job creation, and health co-benefits are calculated at the statewide level.	Health co-benefits, cost savings, and local green jobs should be concentrated in disadvantaged communities.
Public health considerations	Greenhouse gas emissions do not represent a local public health threat. Increased exposure to co-pollutants under cap-and-trade is negligible.	The co-pollutants associated with greenhouse gas emissions are a public health threat to environmental justice communities. Cap-and-trade can create local toxic hot spots.
Public participation	Public participation is limited to comments during the board's quarterly updates. Corporate participation in the cap-and-trade program is prioritized. Western Climate Initiative Inc. is not subject to open meeting or state disclosure laws. The Environmental Justice Advisory Committee is only "advisory."	Public participation in regulatory proceedings should be direct and formative. Cap-and-trade precludes communities from shaping California's climate change law.

of the 11 members of the Environmental Justice Advisory Committee ultimately chose to join a lawsuit against the state. On June 11, 2009, Communities for a Better Environment and the Center on Race, Poverty, and the Environment filed a lawsuit on behalf of 13 environmental justice plaintiffs, collectively referred to as the Association of Irritated Residents (AIR). The litigants claimed that the board's scoping plan violated substantive aspects of AB 32 by failing to study alternatives to cap-and-trade and disregarding the public health of environmental justice communities. They argued that cap-and-trade violated the intent of AB 32 and posed serious implications for communities near polluting industries. According to the plaintiffs' lead attorney, "The harm that our communities will suffer from a poorly made plan will be greater than the harm of not reducing emissions in a way that's responsible, that's legal, and that really reflects the intent and the spirit of AB 32 in the first place." The lawsuit delayed the implementation of the scoping plan for several years.[71]

When the suit reached the trial court hearing on March 18, 2011, the judge, Ernest Goldsmith, dismissed the public health arguments and indicated that the board had legislative authority to interpret the climate change law and had acted "within its discretion, right or wrong, . . . to choose cap-and-trade as the primary methodology." Goldsmith's ruling, however, focused on whether the board had analyzed feasible alternatives, such as a carbon tax, during the environmental review process. He determined that the board sought to "create a *fait accompli* by premature establishment of a cap and trade program before alternatives [could] be exposed to public comment and properly evaluated." The judge held that the board had improperly implemented the plan by starting it before the state's environmental quality review process was complete; as a result, this action undermined public participation. Goldsmith ordered the board to revise the functional equivalent document (an evaluation of the scoping plan's environmental impacts) to include an analysis of alternatives to cap-and-trade. He further instructed the board to halt implementation of the plan until it had conducted an additional public comment period. Most notably, he wrote, "the Scoping Plan fails to provide meaningful information or discussion about the carbon fee (or carbon tax) alternative in the scant two paragraphs devoted to this important alternative. The brief fifteen line reference to the carbon fee alternative consists almost entirely of bare conclusions justifying the cap and trade decision."[72]

The lawsuit angered proponents of cap-and-trade such as Robert Stavins, a Harvard University professor of business and government. He derided the environmental justice lawsuit as "misguided" because AB 32 would not only reduce California's greenhouse gas emissions but also lower the state's overall emissions of co-pollutants. He asserted that if current laws regulating co-pollutants were insufficient, then the best response by environmental justice groups would be "to revisit existing local pollution laws and perhaps make them more stringent."[73] Environmental justice groups dismissed these claims as ignoring a history of lax enforcement of local air quality laws and policymakers' simultaneous reluctance to strengthen them. They further countered that local air pollution laws offered inadequate safeguards for low-income communities of color because regulators only evaluated one pollution source at a time. Evaluations did not include multiple pollution sources and environmental stressors in an area that together could create new, elevated, and unsafe health risks, also known as cumulative impacts.[74]

Former assembly speaker Núñez called the court ruling disappointing and defended the state's neoliberal approach: "It's a false assertion, there won't be more pollution. . . . [This ruling is] another roadblock to California being able to achieve its predominance in the environmental world."[75] Núñez did, however, criticize the board for not taking the concerns of environmental justice groups more seriously in developing the scoping plan. Throughout the two-year legal battle, members of the Latino Caucus remained silent. Without support from their Latina/o allies in the legislature, the court order proved to be a hollow victory for environmental justice advocates. The board produced a supplemental analysis to comply with the judgment but maintained that the cap-and-trade mechanism was the only feasible policy direction. During a 45-day public comment period, the board received 109 comment letters, many from social and environmental justice organizations regarding public health concerns. When the board approved the revised functional equivalent document on August 24, 2011, it argued that the environmental justice comments were irrelevant and referred groups to its public health analyses. The Center on Race, Poverty, and the Environment responded to the revised document with a terse claim that it was a post hoc rationalization of the board's 2008 decision to adopt a cap-and-trade mechanism: "[The board's] supplement and its actions to continue implementing cap-and-trade while creating the alternatives analysis makes a mockery of the letter and spirit of . . . public participation and informed decision-making."[76]

After satisfying the court's ruling, the board unanimously approved the cap-and-trade program on October 20, 2011. The vote made California the first state in the nation to adopt an economy-wide cap-and-trade system to achieve reductions in greenhouse gas emissions. Applauding the board's decision, the former governor Arnold Schwarzenegger declared, "Today's adoption of a cap-and-trade program is a major milestone for California's continued leadership on reducing the world's greenhouse gases. . . . [T]he most critical phase in the fight against climate change is diligently, aggressively, and correctly implementing this law."[77]

Despite the unanimous vote, several board members noted their apprehension about moving the state into uncharted territory and the potential for public health consequences. In response, the board adopted an adaptive management plan that sought to address localized air quality impacts. Adaptive management is a process that promotes flexible decision-making and allows for adjustments in the face of uncertainties, as the results of managerial actions and other events become better understood. The board described this open-ended approach in positive terms, as learning by doing.[78] But environmental justice groups rejected the adaptive management plan, which they called a "trial and error" process. Alegría De La Cruz, legal director with the Center on Race, Poverty, and the Environment, further characterized the plan as unenforceable and too reliant upon the board's discretion: "The adaptive management plan . . . allows for action if [co-pollutant] emissions increases happen. But [the board] has said that *if* it finds there are increases, it has to find that emissions increased as a result of the cap and trade rule. Those causal connections will make it nearly impossible for [the board] to take any action when [co-pollutant] emissions increases happen. Given these two impossibly high hurdles . . . this adaptive management plan will most certainly not address health concerns raised by the cap and trade rule."[79]

With the adoption of the adaptive management plan, California went on to conduct its inaugural cap-and-trade auction on November 19, 2012. Prior to the auction's launch, Núñez again endorsed neoliberal climate action. He stated that California was on the cusp of an extraordinary opportunity to "limit pollution, protect public health, and spur a clean energy revolution." He reaffirmed that in passing AB 32, the legislature had recognized that market-based programs offered a range of environmental and economic benefits. In its first year, California's program generated nearly $1 billion, and it would go on to become the world's second-largest carbon market at the time, trailing only that of the European Union (China would develop a cap-and-trade program in 2017). Cap-and-trade proponents hailed it as a success and proof of the viability of market-based mechanisms.[80]

Who Has the Authority to Create Climate Policy, and Who Benefits?

Environmental justice groups played a leading role in shaping the language of AB 32. Yet the law was implemented in ways that were perceived as disempowering their communities. The California Air Resources Board's decision-making favored abstract, rather than embodied, forms of knowledge and action. The evidence and historical record analyzed in this chapter suggest efforts to incorporate environmental justice into California's climate change regime broke down for two reasons. First, through the adoption of cap-and-trade, climate change was separated from a human and local scale. It was framed as a global problem with no direct public health consequences. Second, in this approach, experts—not activists—held the key to knowledge about both the problem and its solutions. The potential for negative environmental impacts on local communities was largely ignored.

California's climate change regime activates an array of policy instruments, regulations, and agreements, policing boundaries of scale and linking contentious coalitions of actors. It also provokes deeper questions: Who has the authority to create climate change policy, and who benefits from it?[81] Environmental justice groups collaborated with legislators of color based on a shared embodied experience. But this common ground, which formed a basis for political cooperation through complex negotiations, generally did not extend to the board's staff experts or its leadership. When the scoping plan was adopted in 2008, the entire board membership and executive staff was white. By 2011, when cap-and-trade was approved, the board's racial composition included one African American board member and one Asian American on its executive staff. The nature of AB 32's implementation highlights how bureaucratic institutions around climate change science and policy can be highly homogenous in terms of race and class.[82]

Though Latina/o legislators secured equity language in AB 32, they also ceded important implementation decisions to a board whose senior leadership was perceived as not sharing their priorities. These decisions included determining the scale of the scoping plan, the best available research techniques, and the type of interventions for emissions reduction. This discretion empowered the board and allowed its members to maintain their authority as experts and simultaneously to discredit activists' knowledge. Ultimately, environmental justice was institutionalized according to the values of state regulators rather than those of the social movement. Climate change policy in this case was not only shaped by fixed assumptions about credible forms of knowledge but also embedded within complex racial, social, political, and economic structures.[83]

The inability to implement the bill's environmental justice measures was also driven by factors outside the board itself. First was Governor Schwarzenegger's insistence on cap-and-trade, to the point that he undermined the board's authority. Second was the influence of traditional environmental groups who also supported cap-and-trade. Third was a regulatory framework that failed to incorporate the concerns of underresourced environmental justice groups, despite the changing demographics of the state and legislature. Notwithstanding the contentious nature of AB 32, a Latino Caucus member described the passage of the legislation as a critical turning point for people of color taking the lead in the development of equitable environmental policy: "After 2006, everything started to shift. You saw the number of Latino legislators teaming up with environmental justice and traditional environmental groups increasing tenfold—with people of color authoring major climate change and environmental legislation. It opened the doors for us. Before, traditional environmental groups would not go to the urban Latino focused on public health and justice issues. They would always go to the liberal, coastal, white Democrat. I think this was the critical turning point where environmental justice was no longer just a token mentioned in a larger bill but became key to the future drafting and implementation of policy."[84]

Environmental justice advocates further recognized that in the years following AB 32's passage, Latina/o legislators started to consolidate political power as traditional environmental groups sought to be more responsive to their concerns: "At that time, it was obvious, the writing was on the wall—the Latino Caucus was ascending. More Latino legislators would be voted into office, and the old-school white environmentalists were retiring. . . . That gave traditional environmentalists an incentive to build relationships and be responsive and sensitive to what the caucus wanted."[85]

This shift in political power also led to efforts to diversify regulatory bodies, which made them more reflective of the state's new majority. According to another environmental justice advocate, "You had Latino legislators demanding the governor to send them people of color to appoint to powerful policymaking boards and commissions. They wanted to see diversity at the decision-making table. We didn't have a deep political bench, and because of that, it got ugly at times." After Schwarzenegger left office in 2011, his successor, Jerry Brown, was more receptive to Latina/o legislators' demands. He appointed more people of color to influential policymaking positions on the California Air Resources Board, Natural Resources Agency, Energy Commission, and his gubernatorial cabinet.[86]

The contentious and uncertain nature of AB 32's implementation indicates that the formulation of climate change policy is a nascent process. Few standards and professional norms exist to guide efforts. Environmental justice groups have the opportunity to challenge undemocratic planning processes and create new forms of expertise and knowledge. Yet such an endeavor requires multiyear advocacy campaigns, the scaling up of local experiments, and the establishment of diverse coalitions. Even if communities develop "coherent, expert-tested plans that are objectively superior to industry-backed proposals, they will not be adopted without the exercise of political power." Such capability is critical to achieve equitable climate change policy at the local and state levels. California's climate change programs were initially developed through less accountable forums and methods; nevertheless, important lessons emerged from this process. Environmental justice groups and Latina/o legislators raised a critical voice in contestation. This, in turn, fostered opportunities for constituent education, coalition-building with diverse groups throughout the state, and a public record of grievances and deliberations.[87] These actions would provide a strong foundation for the next round of policy development and a new political order for environmental protection throughout California and beyond.

CHAPTER FOUR

Changing the Climate from the Streets of Oakland

> Up until the last decade, climate change tended to be a "White," affluent issue about saving the rainforest and polar bears. It hasn't had deep relevance for people trying every day to get by in the urban centers. Most of the messages of climate change are about eliminating global greenhouse gas emissions and not about people.
>
> —BRIAN BEVERIDGE, codirector of the West Oakland Environmental Indicators Project

IN EARLY 2009, THE Oakland Climate Action Coalition was conceived in the small cinderblock basement of the Ella Baker Center for Human Rights, a nonprofit strategy and action center. The group came together at a discouraging moment for California's environmental justice movement. After years of wrangling between the California Air Resources Board and environmental justice activists over the best methods to implement the state's Global Warming Solutions Act (AB 32), relations between the two had completely broken down. In December 2008, the board had adopted an imple-

Figure 4.1. Flyer for the Oakland Climate Action Coalition youth teach-in. Courtesy of the Oakland Climate Action Coalition

mentation plan that largely ignored environmental justice concerns; by June 2009, environmental justice groups responded with a law suit against the state. In Oakland, however, activists saw a strategic opportunity to shape climate policy at a local level. When the city announced that it would develop a climate action plan, activists took steps to incorporate their perspectives from the streets.[1]

The Ella Baker Center was named after a behind-the-scenes organizer and architect of the civil rights movement who believed in the power of everyday people to change their own lives. Ella Baker was involved with three of the major groups identified with the twentieth-century civil rights movement: the National Association for the Advancement of Colored People, the Southern Christian Leadership Conference, and the Student Nonviolent Coordinating Committee. Over the decades, the center has carried out its namesake's mission and has been at the forefront of supporting better schools, a cleaner environment, and more job opportunities for Oakland's communities of color. So when the city announced its intention to develop a climate action plan, the center was well positioned to act. Rooted in the community, it had the capacity to quickly organize more than 50 people from 30 community-based organizations to outline a comprehensive local climate action plan that addressed the needs of the Oakland residents most impacted by air pollution and poverty.[2]

The first grassroots convening at the Ella Baker Center began with a question: What would a people's energy and climate action plan look like in Oakland? After several hours of dialogue and facilitated discussion, participants covered the basement walls with neon pink, green, and yellow Post-it notes detailing their suggestions and ideas. These included demands for locally produced renewable energy projects and green jobs in disadvantaged communities, affordable housing as a greenhouse gas emissions reduction strategy, and increased public transit options to improve air quality. Through this meeting, the Oakland Climate Action Coalition was officially established. Over time, the Ella Baker Center became the convener of the cross-sector coalition of over 50 organizations that expanded to include environmental and social justice groups, labor unions, green businesses, faith-based organizations, women's reproductive rights groups, and advocates for sustainable development. As the convener, the center provided staff to coordinate the coalition, led the drafting of the mission statement and goals, and facilitated steering committee meetings to ensure that Oakland's climate action plan would comply with the coalition's benchmarks.

The coalition would shape local climate policy in ways that environmental justice advocates at the state level could not. Through a process of collaborative experimentation, the coalition brought embodied knowledge and community concerns into the heart of climate change policy and went far beyond the consultations that took place around the implementation of California's AB 32. In doing so, however, the coalition faced a range of political and conceptual obstacles. As debates around AB 32 revealed, the conceptualization of climate change functions in a network of power relations that can be difficult to shift. A definition of climate change framed in carbon reductionist terms limited the possibilities for community-based solutions. As a result, the places that disadvantaged communities called home were often not seen as part of "the environment," and local problems were smoothed over by statistical methods favoring generalization and large-scale analysis. Growing evidence, moreover, suggests that human health and equity issues are often excluded from climate planning even at a local level. Municipal climate action plans rarely analyze or consider the disproportionate impact of climate change on low-income neighborhoods and communities of color. The disconnect between civic approaches to public health and climate change can leave these communities disengaged from the policymaking process and underrepresented in mitigation and adaptation plans.[3]

Incorporating public health and equity into climate action plans is challenging for several reasons. First, the structural conditions creating environmental health inequities are often obscured by scientific and regulatory approaches that rely on aggregated data. These regional or global variables may not take into account specific risks and impacts to communities from neighborhood-level sources of pollution. Second, a disconnect between climate planning and public health exists because greenhouse gases are considered to have no direct public health impacts. Research by Susanne Moser and Lisa Dilling further suggests that since greenhouse gas emissions are invisible and because we breathe them without getting sick, many policymakers have a difficult time making the links between greenhouse gas emissions, co-pollutants, and public health. Finally, the organizational and conceptual structures through which governments evaluate knowledge and make climate change decisions can undermine attempts at community engagement, even when climate policy explicitly includes environmental justice goals.[4]

Public health and equity can be better served by local climate planning that follows a holistic and collaborative approach. Such an approach acknowledges how climate change is connected to other types of knowledge about the local environment; it also involves different ways of producing knowledge to frame a culture of climate change and its corresponding policy responses.[5] Close examination of Oakland's climate action plan reveals how the city's environmental justice groups used such an approach. They analyzed the effects of climate change through people's local, embodied experience and relationship to community, eschewing sole reliance on expert-produced metrics. In doing so, they transformed the city's civic epistemology—the institutions, conceptual frameworks, values, and practices used in collective decision-making around climate change. Their efforts disrupted conventional policy approaches and produced place-based conceptualizations of climate change that underscored population health and community well-being.[6]

The dominance of experts' quantitative knowledge in climate action planning, however, was evident in the initial stages of the Oakland climate action plan. Conversations with environmental justice advocates helped me understand the prevalence of this approach and how advocates were experimenting to transform top-down technical practice into community-based solutions. According to one advocate, disadvantaged communities viewed the plan as an opportunity to advance environmental justice and local green economic development goals but struggled to persuade city officials to frame it this way: "People get focused just on the greenhouse gas emissions, and they lose sight of the broader community benefits. . . . But working closely with city staff, we were able to reframe it to the city council as a plan that could address global emissions and also public health and equity."[7] In this chapter, I show how open conversations between community members, activists, experts, and city officials generated new kinds of knowledge that changed institutional structures and disrupted the status quo. Through such experimentation, Oakland advocates negotiated the boundaries of scale between the generalized, global perspective and the environmental problems they wanted to address in the streets.

Configuring Oakland's Climate(s)

> Climate change is the greatest public health challenge of this century. . . . [T]he Oakland Climate Action Coalition has done an amazing job in pushing the [City] Council and our communities to imagine and prepare for it.
>
> —DANA GINN PAREDES, training director for Asian Communities for Reproductive Justice

With a population of over 400,000, of whom almost two-thirds are people of color, Oakland has a long and rich history of civil rights and environmental activism. As the birthplace of the Black Panther Party, Oakland has benefited from a culture of self-determination and resistance developed through local experiments to improve the living standards of communities of color. Notable among these are the Free Breakfast for Children Program (which would go on to serve as a national model for public schools) and community health clinics to address issues such as food insecurity and limited access to healthcare. Oakland's culture of capacity-building and experimentation has led its activists to participate in national efforts to disrupt governance practices, deviate from existing rules, and contest sources of authority.[8]

Figure 4.2. Flyer announcing a rally for clean air and climate action. Courtesy of the Oakland Climate Action Coalition

The sophisticated culture of activism in Oakland, like that of Richmond, is also derived from a legacy of inequitable development practices (see Chapter 2). Toxic facility sitings, low socioeconomic status, proximity to one of the nation's busiest container ports, and lack of a fair distribution of environmental goods continue to degrade the built environment in many Oakland neighborhoods. In those neighborhoods, residents face increased exposure to pollution-related public health risks, such as asthma, heart disease, cancer, premature death, and neonatal problems.[9] According to CalEnviroScreen, the environmental health screening tool developed by the California Environmental Protection Agency, more than 50,000 Oakland residents live in neighborhoods listed in the top 20 percent of California census tracts for cumulative environmental impact across the state. These communities sit next to a busy shipping container port, airport, railyards, or freeways. Residents there are exposed daily to far greater levels of multiple forms of pollution from vehicle exhaust and commercial operations than people living in other census tracts.[10] These pollution sources also often contribute to the greenhouse gas emissions that fuel climate change. In pinpointing such hot spots, CalEnviroScreen highlighted Oakland's role in global-local environmental health degradation.

Motivated by these disproportionate environmental burdens, the city of Oakland and local environmental justice groups sought ways to link urban planning, public health, and climate change. They developed an approach that displaced the expert-driven processes that often characterize climate action plans nationwide. Previous research has shown that, similar to the state of California, local governments develop standards and climate policy through the establishment of task forces populated by scientists, university experts, and technical organizations that rarely address public health and equity issues.[11] Bucking these expert-driven norms, Oakland in December 2012 adopted one of California's highest city-scale greenhouse gas emissions reduction targets, following a three-year collaboration with the coalition. In the process, environmental justice groups defined a holistic concept of the environment that identified geographically and socially uneven risks and impacts of climate change and opportunities to promote community-based solutions.

The coalition's strength and success were due to the diversity of its members, who were recruited throughout Oakland's diverse neighborhoods (table 4.1). Together, they were a powerful force that provided multisector expertise on a host of issues, including transportation, affordable housing, energy, urban agriculture, adaptation planning, and community engagement. According to Emily Kirsch, founding coalition coordinator and green jobs organizer for the Ella Baker Center, "Having a diverse coalition with strong expertise is important in these types of policy initiatives. . . . When you talk about climate change, it's food, water, transportation, housing, energy, health equity, and everything you possibly can think of. So we went around to our friends and allies to find out what sort of climate-related projects they were working on and the type of expertise they could bring to the coalition. Then we strategized how we could get these projects included in the climate action plan and on the books as part of the city's plans."[12]

Working with Asian Communities for Reproductive Justice, the

Table 4.1. Oakland Climate Action Coalition membership, 2009–11

Convener		
The Ella Baker Center for Human Rights		
Steering Committee		
Asian Pacific Environmental Network	Communities for a Better Environment	Pacific Institute
Bay Localize	Local Clean Energy Alliance	Transform
Causa Justa: Just Cause	Mobilization for Climate Justice West	Urban Habitat
Center for Progressive Action	Movement Generation Pacific Institute	West Oakland Environmental Indicators Project
Coalition members		
Asian Communities for Reproductive Justice	Greenbelt Alliance	Progressive Jewish Alliance
California Food and Justice Coalition	Greenlining Institute	Star King School for the Ministry
DIG Cooperative	Hope Collaborative	Unitarian Universalist Legislative Ministry California
Environmental Defense Fund	International Longshore and Warehouse Union Local 6	The Workforce Collaborative
First Unitarian Church of Oakland	Kehilla Community Synagogue	
Forward Together	Mandela Marketplace	
	Planting Justice	
Allies		
Alameda County Green Business Program	Food First/Institute for Food and Development Policy	Pesticide Watch
Alliance for Oakland Development	Global Alliance for Incinerator Alternatives	Pueblo
America Works	Greywater Action	Rising Sun Energy Center
California Interfaith Power and Light	Grind for the Green	Sierra Club
CiviCorps	International Brotherhood of Electrical Workers Union Local 595	Sungevity
ClimateChangeEducation.org	Live Real	Sun Light & Power
East Bay Alliance for a Sustainable Economy	Northgate Environmental Management	Sustainable Earth Initiative
Environmental Conservation Options	Oakland Food Connection	Sustainable Economies Law Center
	Oakland Resilience Alliance	US Green Building Council
		Wellstone Democratic Club

Source: Oakland Climate Action Coalition

coalition also built an intersectional campaign around the links between women's reproductive justice and climate change. They held organizing and educational events questioning how the presence of toxic chemicals and greenhouse gases harm reproductive health and also contribute to climate change. The coalition specifically focused on a holistic life-cycle analysis of chemicals and how pollution exposure in nail salons and electronics factories is impacting the natural environment and women's bodies.[13]

The coalition's broad range of expertise and priorities, moreover, enabled it to work with the city to move the climate action plan beyond just technical metrics to a community-based plan. In developing the framework for the plan, Oakland officials had initially followed conventional methods. Officials identified greenhouse gas emissions reduction targets, environmental priority areas, and strategies to address targets by consulting with experts at nongovernmental organizations such as ICLEI–Local Governments for Sustainability and the private consulting firm Circlepoint Inc.[14] The city also adopted standard approaches to public participation that environmental justice groups saw as too "top-down."

The initial city workshops (attended by around 200 people representing the coalition, government agencies, utilities, interest groups, businesses, and individual residents) focused on informing the community about the methods that city staff had selected to develop the plan. Early in the process, however, the coalition approached the city and requested more direct involvement. The coalition asked the city not to establish a formal expert task force and to instead allow a community-based approach. The coalition included long-established community members; due in part to their political influence, the city council ultimately allowed the coalition to facilitate and fund a parallel community advisory process. The city's comparatively small scale (as opposed to the state) made this strategic opportunity possible, through personal connections and closer interaction between influential community members and policymakers. As a coalition member recalled, "The city did host their own workshops, but they are pretty boring and held at 2 p.m. . . . We attended and gave our input. That is because we get paid to attend. But we wanted to hold workshops that were more accessible to the public and were fun and engaging. . . . So we hosted a series of workshops in the flatlands of East and West Oakland, knowing that communities most impacted by climate change are often least represented in terms of decision-making."[15]

Through this collaboration with the coalition, Oakland adopted one of California's highest municipal greenhouse gas reduction targets: a 36 percent reduction from 2005 levels by 2020 and an 85 percent reduction by 2050. Although other California cities have also set such goals, Oakland's target levels are some of the first to comply with the recommendations of the Intergovernmental Panel on Climate Change. Oakland's targets also surpass California's statewide requirement to reduce carbon emissions to 1990 levels by 2020 and are more than double the state's recommendation that local governments reduce emissions 15 percent by 2020.

The coalition strategically pushed for higher reduction targets as leverage for additional measures to address their equity concerns. One coalition member recalled, "We knew if we pushed for a high greenhouse gas target, the more comprehensive the plan had to be. And it could include measures that were community-based, in addition to the standard greenhouse gas mitigation solutions."[16] As a result, Oakland is also one of the first cities to explicitly link evaluative criteria with benefits for disadvantaged communities when weighing climate policy choices (table 4.2). In this process, the plan's authors considered whether its benefits outweighed the burdens on disadvantaged communities. For example, they worked to preserve affordable housing in transit-oriented development projects, in a bid to ensure that this greenhouse gas emissions reduction measure would not displace low-income residents.[17]

The coalition argued that transit-oriented development projects (that is, housing, retail, and office uses located next to public transit) in existing high-density neighborhoods could be an effective greenhouse gas mitigation measure, but resulting displacement of low-income people, senior citizens, and renters could undercut mitigation goals. Lower-income residents might be forced out to cheaper suburbs with fewer transit options if their neighborhood's older housing stock was replaced with new market-rate units. Many individuals might have to buy a car to commute to work and access community services, thereby increasing the region's vehicle miles

Table 4.2. Oakland's evaluative criteria for potential energy and climate actions

Evaluative criteria	Issues to consider
Greenhouse gas reduction potential	Magnitude of greenhouse gas reductions Measurability of reductions
Implementation cost and access to funding	Cost to city budget Cost to other stakeholders Access to funding
Financial rate of return	Return on investment to city and/or stakeholders in implementing the actions Protection from future costs
Greenhouse gas reduction cost-effectiveness	Relative cost-benefit assessment in terms of estimated greenhouse gas reductions
Economic development potential	Job creation potential Business development and retention potential Workforce development potential Cost savings for community
Creation of significant social equity benefits	Benefits to disadvantaged residents in the form of jobs, cost savings, and other opportunities Reduction of pollution in heavily impacted neighborhoods Equity in protection from impacts of climate change
Feasibility and speed of implementation	Degree of city control to implement the action Level of staff effort required Resources required Degree of stakeholder support Amount of time needed to complete implementation Time period during which implementation can begin
Leveraging partnership	Leverage partnerships with community stakeholders Leverage partnerships on a regional, state, or national level Facilitate replication in other communities
Longevity of benefits	Persistence of benefits over time Opportunity to support future additional benefits

Source: Adapted from City of Oakland, "Energy and Climate Action Plan," appendix

traveled and its greenhouse gas emissions. Oakland was the nation's first city to link climate change policy with affordable housing in this manner. The coalition also fought to include neighborhood-scale adaptation planning in the climate action plan to address the most harmful near-term effects of climate change on socially vulnerable communities. In contrast, conventional adaptation studies at the time often focused only on protecting hard assets, such as vital city infrastructure, or ecological systems.

In addition to its role in promoting community-based policies, the coalition was key in the overall framing of the climate action plan. Garrett Fitzgerald, Oakland's city sustainability coordinator, praised its efforts: "The [coalition] made my job a lot easier by providing smart, specific recommendations for the plan and doing a lot of work to bring more of Oakland's voices into the process. It's rare to find community partners as dedicated and willing to collaborate with city staff as the [coalition]."[18] Even before city staff released their first draft of the climate action plan, the coalition had already developed and presented its own comprehensive plan to city officials, based on the community workshops they had hosted. According to a member of the coalition's steering committee, "What drew these unlikely partners together is the goal of a just and equitable energy and climate plan for the city. Whether they were a green enterprise looking to grow their business in a green and sustainable way, or a labor union looking to ensure jobs in a new economy for their members, or an environmental group that has done the research to know the catastrophic effects of global warming—they all had a stake in making sure that the plan was done right for the city of Oakland."[19]

Through this parallel policy development process, the coalition produced 50 of the 150 greenhouse gas reduction measures and goals in the final version of the climate action plan that the city council adopted. The coalition's policy committees used research and embodied knowledge to build justifications for specific greenhouse gas reduction measures and public health targets.[20] These committees addressed areas such as transportation, affordable housing, and land use; building and energy use; consumption and solid waste; food, water, and urban agriculture/forestry; and adaptation, resilience, and community engagement. The committees convened several times a month and were led by two cochairs—one from a policy-based organization and the other from a grassroots group—to balance expertise in policy development with on-the-ground experience in their recommendations.[21]

Many municipalities acknowledge that public participation should play a role in formulating climate change risks and strategies. However, officials frequently claim they cannot garner significant public interest because the science is complex and climate change is a long-term and uncertain process. As a result, many cities opt to establish an expert task force and hire environmental consultants instead.[22] Compared to most municipal climate policy planning, Oakland represents an innovative case in six key ways:

1. It included local, embodied knowledge in the development of climate policy.
2. Public participation was embedded in the regulatory science and policy processes.
3. Its understanding of climate risks and impacts focused on the human scale.
4. Measures were chosen for their potential health benefits.
5. Adaptation plans focused on socially vulnerable communities.
6. The climate action plan included explicit references to equity and environmental justice (table 4.3).[23]

The coalition's climate policy development process disrupted conventional climate planning by reducing the primacy of scientific advisors and validating embodied knowledge held by communities. Oakland's climate action plan is an innovative case, moreover, because the coalition is officially listed as a major contributor to the plan's development. This is a rare occurrence. As a long-serving member of the Oakland City Council noted, city collaboration with the coalition produced a plan that stands in sharp contrast to previous environmental documents produced by the city: "I've been a Council member for 16 years and I've seen a lot of environmental plans. Oakland's Energy and Climate Action Plan is unique because it lifts the voices of low-income communities and communities of color."[24]

The work of the coalition to develop, pass, and implement the

Table 4.3. Climate action plan metrics in California

	Conventional climate action plan	Oakland climate action plan
Policy processes	Expert knowledge emphasized	Local, embodied knowledge valued
Public participation	Top-down approach	Ground-up approach
Scale of climate impacts	Exclusive focus on hard assets at the city scale	Inclusion of the human scale (i.e., neighborhoods)
Co-benefits	Cost savings, efficiency, economic development	Public health, cost savings, efficiency, local green jobs and energy
Focus of adaptation	A normative goal focused on citywide infrastructure and ecological systems	Focus on socially vulnerable populations/neighborhoods, citywide infrastructure, and ecological systems
Social equity or environmental justice	Rarely cited in documents	Cited as a guiding principle of policy development and implementation

city's climate action plan makes Oakland a model for what communities across the country can do to shape their climate change policy to local needs. Its direct engagement represents an experiment in how local climate governance can be established and defined. First, the plan focused on socially vulnerable communities with the most to lose from the impacts of climate change and suggested neighborhood adaptation plans, not only mitigation measures. Second, the coalition helped create a climate policy model based not just on greenhouse gas reductions but also on ensuring multiple community benefits in the form of green jobs, affordable housing, and health co-benefits. Finally, it brought together a diverse community and created a transformative space to challenge the power relations around climate change policy at the local level.

Localizing Climate Change for Community Action

The coalition influenced climate knowledge, thus transforming a global phenomenon into a local one with practical applications. It did so through community engagement, political mobilization, education, and an experimental approach. Taken together, these exceeded the formalities of neocommunitarianism (as in the case of the state's AB 32 implementation) to involve members of the public in identifying local problems associated with climate change and in finding equitable solutions. The coalition's work with residents—in particular, low-income families and communities of color—was highly popular and inclusive. This was a key factor in the adoption of the coalition's recommendations in the final plan.[25]

To transform how climate change was perceived in Oakland, the coalition convened and funded 14 workshops. These workshops, along with convenings and rallies, engaged more than 1,500 residents to develop local solutions to climate change, compared to the 200 individuals who attended the city-sponsored events. Several workshops were conducted in multiple languages—for example, the nonprofits Movement Generation and the Asian Pacific Environmental Network (which has offices in Richmond and Oakland) facilitated Spanish- and Chinese-language workshops for immigrant residents. The inclusive process produced widespread support for and engagement with the plan by Oakland residents most impacted by pollution and poverty.

The coalition used youth engagement programs to further localize climate knowledge. For example, the coalition hosted a solar-powered concert, featuring legendary hip-hop artists Pete Rock and C. L. Smooth, to promote a Climate Adaptation Work Day at Laney Community College. More than 350 Oakland residents, many of them youth, helped install a garden and rainwater catchment system at the college. Coalition member organization Forward Together organized 80 East Oakland high school students in role-playing activities designed to envision what climate solutions in their homes, schools, and neighborhoods could look like. A Community Convergence for Climate Action was also held to showcase a theatrical performance by high school girls on climate change, live hip-hop concerts, and a report-back session from residents who had attended the coalition's climate workshops. The Community Convergence event demonstrated the high level of interest from local residents in developing the climate action plan and created a space for them to articulate climate solutions that would make a real difference in their own lives.[26]

The coalition also facilitated workshops on disaster preparedness for low-income communities that focused on the risks and impacts of climate change through interactive games and learning initiatives. These included the *Are You a Climate Change Survivor?* activity workbook; board games like Climate Justice Human Bingo and Community Resilience Lifeboat, and fact sheets with activities designed to raise awareness about climate change impacts.[27] Through such collaborative projects, Oakland set the trend for a holistic approach to climate action planning. These activities aimed to help diverse people and organizations imagine and implement solutions to protect residents from the localized threats of climate change: heat waves, floods, wildfires, poor air quality, and rising utility costs. Brian Beveridge, co-executive director of the West Oakland Environmental Indicators Project and coalition member, noted, "We started by bringing people together and talking about assets and vulnerability—talking about things they want to protect. It starts as a mapping exercise; we look at all the places we are strong before we look at our vulnerabilities. . . . At the community level, it is not technocratic. You can't just say there is some technological fix for people, because we are really not protecting hard assets; we are talking about people surviving as a community during a disaster."[28]

In addition to community engagement events, the coalition turned to electoral politics—the 2010 mayoral and city council races. On March 30, 2010, the coalition organized a 200-person rally at city hall, where Oakland City Council members and candidates listened to labor and green business leaders' recommendations for the climate action plan. Andreas Cluver, secretary-treasurer for the Alameda County Building Trades Council of labor unions, described a developing relationship among climate change, local politics, and community interests: "I spoke at the [city hall] rally to show that labor leaders and community leaders are united for job-creating climate solutions. By passing a strong [climate action] plan, we can get our members off the bench and into jobs."[29] The coalition also hosted a larger formal event, the Green Mayoral Forum on September 14, 2010, where more than 2,000 local residents—a majority of them people of color—convened to listen to the candidates' proposals for how they would advocate for the adoption and implementation of the climate action plan. Hosting the Green Mayoral Forum and the city hall rally set a strong precedent, which showed that city elected officials must develop an explicit policy agenda linking climate change with direct benefits to their communities.[30]

Participatory Research in Climate Adaptation Planning

Climate action planning is often focused on the methods that cities can use to mitigate greenhouse gas emissions. But the reality is that climate impacts are happening now and rapidly increasing. Understanding how communities and individuals can adapt to and prepare for those changes is important. However, due to lack of city resources and expertise, adaptation planning was at first given only a cursory mention in Oakland's climate action plan. As an environmental justice advocate noted, the coalition had to develop new methods to analyze neighborhood-scale climate impacts and safeguard socially vulnerable populations: "As originally drafted, the adaptation component of the [climate action plan] was just a text paragraph and basically stated that, yes, this is an issue that the city should be looking at in the future. There were no action measures associated with it in the plan. I would say adaptation planning for socially vulnerable communities is a hard sell, especially when there are limited resources to plan and protect the entire city. We didn't see anyone talking about what happens to communities like West Oakland if we can't avert climate degradation. How do these neighborhoods adapt? So we, along with the Pacific Institute, started this whole discussion of adaptation, vulnerability, and resilience."[31]

The coalition, through grants from the California Energy Commission and the San Francisco Foundation, developed its own local adaptation planning data and models. These would supplement the climate action plan and help policymakers and affected communities decide where to focus climate change programs and resources. Through a community-based participatory research process, they identified geographic areas within Oakland with heightened risks for projected climate impacts.[32] The Pacific Institute, a research think tank and coalition member, administered the adaptation study to involve affected residents in developing solutions that crossed conventional boundaries and ideas. The institute convened adaptation research meetings from August 2010 to November 2011, in collaboration with coalition members at key points in the research decision-making process. These decisions included the types of climate impacts and vulnerability factors to be considered, the interpretations of research results, and ways to share the results with key audiences.[33]

The coalition's chosen methodology included, first, obtaining geographic data on the projected physical impacts of climate change to determine exposure; second, selecting data on indicators of social vulnerability that relate to these impacts at a neighborhood scale; and third, overlaying vulnerability and exposure layers to produce a composite map. According to the project's lead researcher, Catalina Garzón, the adaptation planning process placed a strong emphasis on local context and engaging community stakeholders in defining priorities and solutions: "We held a total of six joint sessions with coalition members, called 'research report-backs,' at key decision points in the research process. We presented research and draft methods for best practices in the field for looking at community vulnerability in adaptation planning. . . . We facilitated discussion about the implications for coalition members' work on local climate change—to understand what would be useful for their own work. Then we would present draft results from the mapping and climate models and asked members what adaptation measures can be best implemented locally to address climate impacts and social equity."[34]

This collaborative process enabled the coalition to focus on detailed models that analyzed city and neighborhood vulnerability to a number of climate impacts, such as coastal flooding due to sea level rise and storm surges, extreme heat, wildfires, and increased particulate matter concentrations from warmer temperatures. Through these models, the coalition argued that Oakland's most socially vulnerable populations were less likely to prepare for, respond to, and recover from these projected threats. Nearly half of Oakland's residents live in areas of high to medium social vulnerability to extreme heat, flooding, poor air quality, and wildfires (fig. 4.3). However, the areas of highest social vulnerability are concentrated in the low-lying areas near the Port of Oakland, major freeways, and the Oak-

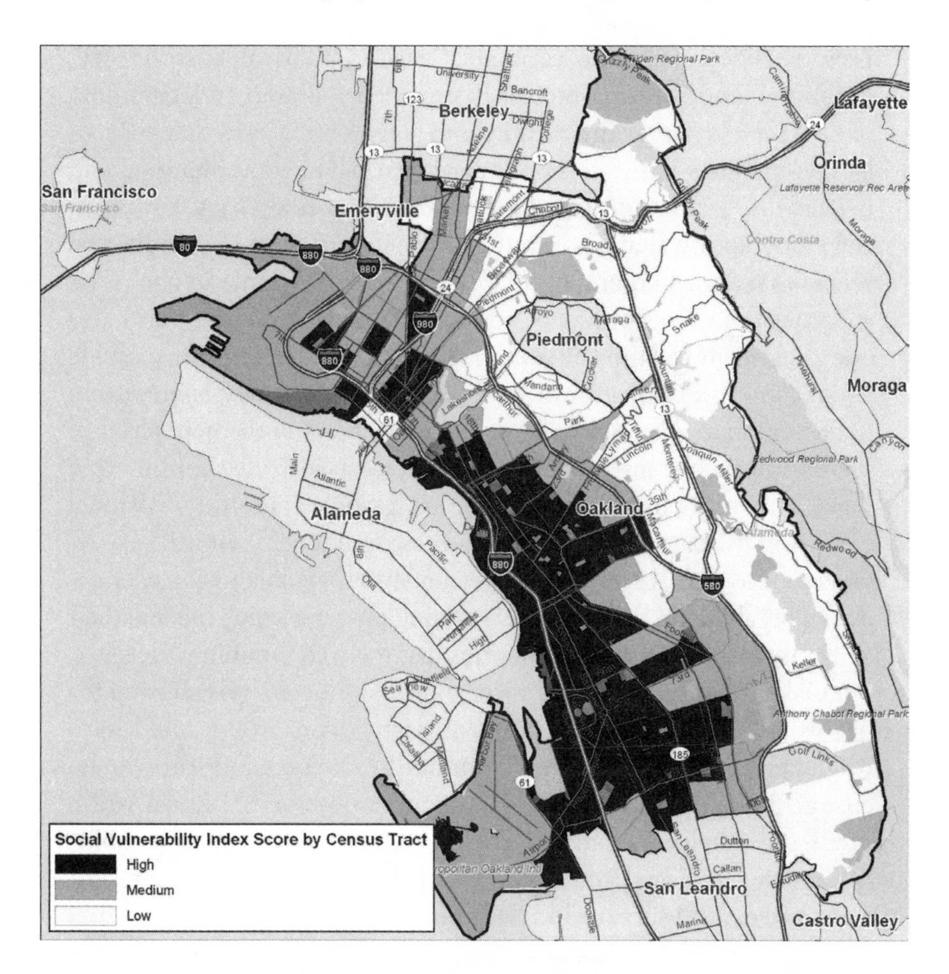

Figure 4.3. Social vulnerability and climate change in Oakland. Courtesy of Garzón et al., *Community-Based Climate Adaptation Planning*

land International Airport. When the research results were completed, the coalition held a final workshop focused on equity and community resilience on November 11, 2011. It attracted more than 100 participants. Pacific Institute researchers presented the findings, and participants brainstormed about their greatest concerns and potential policy solutions at breakout sessions.[35]

The coalition utilized some of the newest approaches to social vulnerability and adaptation modeling, which allowed it to identify the potential impacts of climate change on population groups connected to particular places and communities. This multidimensional view considered climate variability and its political, institutional, economic, and social contexts. It provided a better understanding of how these dynamic interactions can heighten the risks that particular communities face. Moreover, residents' values and opinions about needed protections informed the process. According to Al Irwin, social vulnerability analysis requires an examination of how individual residents and their communities view the world: "Judgments about risk and safety will reflect one's position in the social structure—and also one's degree of trust in the social institutions which currently decide about these questions on others' behalf." A full analysis entails an understanding not only of how people interact with their physical surroundings but also of how they interact with institutions that make environmental policy decisions.[36] Such a multidimensional view is often absent from adaptation research. Facing limited expertise and resources, many local governments are unable to consider social vulnerability in their climate planning process, which hinders their ability to produce equitable solutions that meet communities' needs.

Oakland's experiment in social vulnerability analysis allowed for negotiation among a wide range of participants and granted local residents from disadvantaged communities a voice in local climate change planning. As one coalition member observed, the adaptation process also helped serve as a basis to engage community stakeholders in other regional and statewide projects to address climate impacts and community health: "Through the coalition's advocacy and research, I think we succeeded in getting more focus on social equity in the plan. . . . We are now focused on implementation—ensuring that adaptation planning is included in all land use projects and approvals. . . . We are now working on convening an interagency working group at the regional level, including the Alameda County Public Health Department, Bay [Area] Air Quality Management District, and [the state's San Francisco] Bay Conservation [and] Development Commission to develop a strategy with local community stakeholders to address adaptation."[37] The coalition's participation, moreover, helped improve research design by refining research questions, methods, and instruments for greater accuracy and relevance. Most significantly, the coalition's participation in the data analysis enhanced the interpretive validity of the research findings relating to social vulnerability and climate adaptation.[38]

Civic Epistemologies and Oakland's Transformative Climate

Oakland's climate action plan became a site of innovation in both the production of knowledge and the ordering of political activity. The plan was more than just a technical development in how a locality measures and tracks climate change; it represents new experiments in environmental governance. Such experiments, according to the science policy scholar Clark Miller, "are important features of new emerging civic epistemologies in local, regional, and global settings. . . . [T]hey are technologies through which people are co-producing new ways of knowing and ordering the world at these scales."[39] The construction of climate policy in Oakland resulted from extensive experimentation with public participation in expert advisory processes. This open-ended, inclusive form of consultation gave rise to new local civic epistemologies and a conceptual model of climate change that focused attention on disadvantaged communities.

Governments, as producers and consumers of knowledge, hold an important space in the development of civic epistemologies. Governments often define environmental issues, their scale, and the terms in which residents are included in relevant policy decisions and debates. This process influences which knowledge claims are more likely to be considered valid and used in environmental governance (see Chapter 3). Governments also use quantitative statistics as an instrument of statecraft "to imagine society, the economy, and the nation and to lend to the exercise of public policy a semblance of rationality, control, and accountability."[40] The result is a tendency to privilege large-scale, scientific, and generalized ways of understanding and to deny credibility to embodied knowledge in the context of policymaking and public debate. In Oakland, the coalition successfully argued for the equal relevance of other kinds of knowledge in climate policy and governance. In this manner, it created a form of consultation that earned real trust from local residents, addressed their concerns, and engaged their creativity in imagining ways to combat the risks they face from climate change.

Defining both local and global climate change depends on how knowledge is produced and on the relationships of power among individuals, groups, and institutions in society (table 4.4). In Oak-

Table 4.4. Oakland's civic epistemology of climate change

	Values, definitions, and practices
Spatial frame	The spatial frame is **multiscalar**, encompassing local and regional, as well as statewide and global, settings.
Climate goals	Climate goals are set in relation to **co-benefits potential**, linking greenhouse gas reductions with co-pollutants reductions and other community benefits.
Forms of knowledge	**Embodied knowledge** is valued alongside professional expertise.
Institutional governance	**Community-based organizations** collaborate actively with government agencies and businesses.
Definition of climate change	Climate change is defined inclusively, with reference to the urban environment, ecosystems, adaptation and greenhouse gas mitigation, **co-pollutants, public health, economic opportunity, and equity**.

Source: Adapted from Miller, "New Civic Epistemologies of Quantification," 409

land, embodied knowledge entered meaningful dialogue with technical practices to overcome the problems that arise when global policy approaches are introduced into local contexts. The contrast in approaches to consultation—and the outcomes that followed—between the implementation of the state's AB 32 and the formulation of Oakland's climate action plan demonstrates that "global solutions to environmental governance cannot realistically be contemplated without at the same time finding opportunities for local self-expression."[41]

This insight and these forms of activism have pressured experts and policymakers to relinquish some control over how climate change is defined as a public concern and how research on solutions is formulated and conducted. In Oakland, climate policies have become intertwined with attempts to reimagine what the city stands for and how environmental justice is achieved. In responding to global climate change, activists assert their local roots and exercise a degree of voice, representation, and agency.[42] Oakland's climate action plan, moreover, introduced an innovative civic epistemology of climate change that encouraged genuine collaboration between regulators and environmental justice groups. While its actors gathered, evaluated, and used scientific and public health knowledge in fundamentally different ways, they were able to work together to define the problem and determine which climate risks required intervention. Oakland's environmental justice communities assumed a central role in the city's climate change interventions. They earned valuable recognition for their understanding of socio-ecological dynamics and for applying that knowledge to the policymaking process to benefit their communities. Oakland's experience represents a critical model for the future.

Climate Change and Community Well-Being

The Oakland climate action plan represents power and scalar shifts in climate change governance. The coalition's engagement changed the location of authority, as community advocates ruptured experts' exclusive control over the definition and production of climate knowledge. Experimentation in Oakland demonstrates the value of including different forms of knowledge in producing climate change solutions. But such strategies have limitations and are not easily replicable. Communities that have less social capital to expend, that lack a history of activism, or that have fewer resources to organize and develop knowledge frameworks would be less likely to benefit from the same kind of opportunities.

As a new analytical and policy domain, local climate change poses challenges for scientists, city planners, residents, and policymakers. The combination of emerging scientific methods and competing policy goals requires approaches that acknowledge the difficulty of localizing global climate change science while maintaining technical legitimacy and social authority. Initiatives in Oakland highlight a novel approach. They value and reflect embodied experience, culture, social practices, and vernacular understanding, all of which are quite different from knowledge produced through lab experiments or disciplined empirical observations.[43] Oakland's climate action plan, in essence, is a representation of the natural world that gained validity through a local, mutually sustaining interaction between the physical world as described by science and the moral and political world in which local values operate. Sheila Jasanoff elaborates that conventional policies "detach global fact from local value, projecting a new totalizing image of the world as it is, without regard for the layered investments that societies have made in worlds as they wish them to be."[44] As the case of Oakland demonstrates, the claims of climate change science are trusted more when technical practices, cultural values, and democratic politics are more closely integrated. The conceptualization of climate change in Oakland exemplifies the emergence of embodied knowledge as a resource for achieving greenhouse gas reductions, neighborhood scale adaptation, green jobs, health benefits, and community well-being. Through a civic epistemology that operates on multiple scales, Oakland's definition of "expert" has expanded to include low-income communities of color. Environmental justice groups are generating new kinds of data for the localization of climate change policy, calling critical attention to the cultural and human dimension of knowledge production and local practice.

At the scale of the city, environmental inequalities are often less easily smoothed over by technical generalizations, and the complexities of life at the human scale cannot be ignored. In the process of developing Oakland's climate action plan, the relatively small scale of municipal government and a history of activism made it possible for coalition members to substantively engage with policymakers and create space for themselves to collaborate with the city from the beginning. However, it has been far more difficult for environmental justice advocates to engage with policymaking at the state level; the relationships and resources were initially lacking, and the structure and scale of California's regulatory agencies can often limit opportunities for meaningful participation. Scaling up environmental justice strategies came to represent a significant challenge for activists in Oakland, Richmond, and throughout California. They were confronted with apparently insurmountable odds as they turned their attention to transforming the state's regulatory culture and guaranteeing that their local climate change experiments could make a difference in the wider world.

CHAPTER FIVE
Cap and Trade-Offs

Like the Oakland activists who helped develop the city's climate action plan, the broader environmental justice movement tends to focus its concerns around disputes attached to local places and identities. Their efforts to combat global climate change, however, complicate those values. In responding to the development of global carbon markets, environmental justice groups employed new institutional practices that enabled activists to move between scales while challenging the spatial and social implications of climate change solutions.[1]

During my time in the capitol, I saw firsthand a confluence of factors pressuring environmental justice groups to uproot and expand the scale of their local advocacy efforts. This included failed attempts by multinational oil companies, such as the Chevron Corporation, to introduce state legislation that would bypass local environmental review for expanding refinery operations. Such proposed laws would have made it easier for them to transport oil sands from Indigenous lands in Canada and North Dakota to refineries near disadvantaged communities in Richmond and Oakland. This globally connected resource extraction process would not only increase emissions in California but also could have detrimental effects for Indigenous communities throughout the continent. Nonetheless, the anxiety over the unforeseen local impacts of the Global Warming Solutions Act (AB 32) seemed to be the key motivator for environmental justice groups' increased lobbying presence at the capitol. By 2009, several organizations had sent local staff to Sacramento to learn to be state-level advocates. Almost immediately, these rookie lobbyists reframed the debate in climate policy committee hearings. They focused their advocacy on a central question: Who pays for, who crafts, and who benefits from California's climate policy? They argued passionately with legislators that while California's cap-and-trade system held the promise of reducing emissions at a low cost to polluters, it failed to address the underlying equity issues that shaped disadvantaged communities' embodied experiences of climate change. These encounters offered firsthand evidence of how environmental justice groups were engaging in climate change from the streets—a critical reevaluation of both the practice and politics of reducing carbon emissions.

Introducing a view from the streets into California's climate change programs, however, was a contentious endeavor. Such attempts revealed deep divisions between environmental justice groups opposed in principle to using market-based mechanisms as solutions to inequity, on the one hand, and state regulators who focused on cost efficiency and scientific rigor, on the other. Pragmatically, though, some environmental justice groups decided to participate in new efforts to link global carbon markets with local concerns. In California, the central element of this rescaling activity became the Climate Change Community Benefits Fund.[2] Through this fund, the state invests a significant portion of cap-and-trade auction revenue in communities most affected by air pollution. The climate fund supports local solutions that jointly reduce global greenhouse gases and co-pollutant emissions, such as transit-oriented development, renewable energy, and urban forestry projects. It also seeks to create green-collar jobs in communities with high unemployment.

The establishment of the climate fund shows that environmental justice groups' opposition to cap-and-trade was not an outright rejection of the goals of AB 32. Rather, it was a demand that climate change solutions produce more equitable outcomes for all communities. This chapter discusses the multiscalar strategies and scientific methods that environmental justice groups employed to transform the regulatory culture around California's climate change policy. It provides an in-depth analysis of how, after four years of lobbying and a gubernatorial veto, environmental justice groups successfully pushed for the creation of the nation's first Climate Change Community Benefits Fund. Through such efforts, some of the same environmental justice groups whose experimental approach succeeded so well in Oakland began experimenting again—this time, with new tools to institutionalize distributive and procedural justice in climate change policy at the state level.[3]

Transforming the Regulatory Culture of Climate Change

Motivated by their success in creating a more equitable Oakland climate action plan, the Ella Baker Center for Human Rights and the Asian Pacific Environmental Network together sought new strategies to further their local work. They asked how climate change could invite new ways of thinking and doing that would simultaneously promote economic growth and protect the environment. They saw the advancement of a climate fund as an opportunity to link their work on community-based solutions to the state level and offer models for equitable climate change mitigation and adaptation projects. Emily Kirsch, the center's green jobs organizer, told me that when she found out that billions of dollars could be generated from polluting industries under cap-and-trade, she quickly realized that potentially none of it would go to disadvantaged communities. There were no mandates or consultation with local communities about how best to invest the revenue. She saw the climate fund as an opportunity to point to cities such as Oakland and proclaim, "That's where this money should go, to implement climate solutions rooted in equity and projects focused on green collar jobs." Kirsch further stressed that the high degree of inequality within California makes linking local equity concerns with state-level climate change policy design important. Efforts such as the climate fund, she said, could connect poverty alleviation and pollution mitigation. For example, spending on solar power and energy efficiency programs for low-income residents would clean up the environment, reduce residents' utility bills, and create local jobs.[4]

Mari Rose Taruc, the statewide organizing director for the Asian Pacific Environmental Network, also felt a strong responsibility to use the lessons learned in Oakland to influence larger statewide policy. While she believed that Oakland had adopted a strong climate action plan, she could also see the lack of funding as the biggest barrier to implementation: "A lot of cities have climate action plans that are grand ideas, . . . but how are they going to get the resources to support the implementation or creation of the infrastructure for those kinds of ideas?" For her, direct, state-level lobbying was important to ensure that funds would flow back to local communities and that California's global carbon market would have real benefits for environmental justice communities.[5]

To the Oakland advocates, while it was clear that their climate action plan gave them the type of formal authority needed to establish equitable solutions, that authority had to be made real in practice—through action and funding at various scales and through multiple institutions. To achieve this, they had to transform California's regulatory culture of climate change. Environmental justice advocates rose to this challenge by reconfiguring ideas, resources, and relationships and by persuading people in influential state-level regulatory and legislative roles to engage with them in implementing experimental policy. According to the political scientists Margaret Keck and Rebecca Abers, the transformation of regulatory institutions is a non-zero-sum struggle for "practical authority": the problem-solving capabilities and recognition from key decision makers that allow community members and activists to change regulatory behaviors and actions.[6]

The political compromises required to pass AB 32 were contentious and left many unresolved ambiguities that environmental justice groups had to address outside the legislative process. The bill mandated significant statewide reductions in greenhouse gas emissions; it also required California to ensure benefits at the neighborhood scale and to consider the potential impact of reduction measures on communities already burdened by air pollution. But the legislation did not specify how any of these aims would be achieved. Implementation efforts revealed vast differences between how state regulators and community advocates defined environmental justice and climate action.

Personal and institutional responses to issues such as climate change develop through often-conflicting goals and values. Costs and benefits are subject to cultural definitions that frequently occur in the very process of policy design and regulatory implementation. Successful navigation of these processes demands "attention [to] how the rules of the game are set, how winning and losing are defined (i.e., how actors come to understand their interests), and on how such definitions change through social processes."[7] Hence, as Janelle Knox-Hayes contends, the legitimacy of climate policy is also based on the interaction of three types of authority: scientific, developed through technical expertise; economic, created through bureaucratic institutions and economic instruments; and emotive, developed through public response based on feelings and values. All these elements are essential for designing functional climate change policy seen as legitimate by a diverse range of interested parties. Market-based governance, thus, is not a value-neutral process but one that invokes sociopolitical outcomes embedded within cultural contexts.[8] While California has one of the world's most sophisticated and ambitious cap-and-trade programs, fierce debates continue over how—and whether—it can achieve goals related to environmental justice.

In their day-to-day work, environmental justice advocates negotiate complex institutions and norms. Keck and Abers further argue that these advocates change ideas, resources, and relationships by building coalitions and networks of allies and by problem-solving through experimentation. They gain practical authority through these interrelated activities, although this is always a complex and contingent process that various factors can enable or hinder. For example, the implementation of environmental justice values in AB 32 met significant opposition. Activist groups could not align their equity goals with the conventions of scientific evidence required by regulatory agencies. They lacked strong engagement practices necessary to transform ideas, resources, and relationships into recognition of their community-based climate solutions. Without such capabilities, they had limited opportunity to establish their practical authority within AB 32's implementation process.[9]

Successful experimentation with new climate change solutions often begins through collaborative efforts at the local scale that focus on enacting achievable goals (for example, the Oakland climate action plan). In the process, activist coalitions strengthen existing relationships with regulators and develop new skills and knowledge as they do so. Such experimentation can be scaled up if others perceive their efforts as valuable. This recognition can transform the bounds of possibility, making ambitious and formerly impractical ideals seem feasible. Through new ways of thinking and doing, conflict between the economic, social, and cultural dimensions of climate change can be resolved in practice.[10]

The case of AB 32's implementation reveals that carbon markets are more than a mechanism for trading emissions reduction credits. They represent networks of actors that are embedded within and are extensions of social, political, and cultural institutions. Carbon markets are constructed through a global set of ideals, but they are also unique expressions of the places in which they operate.[11] Through new forms of practical authority, I argue, California environmental justice groups are engaging regulators, legislators, and other experts to help transform the local expressions of the state's climate change program.

Toward a Legislative Redress of Carbon Markets

Throughout the AB 32 implementation process, environmental justice groups questioned policymakers on how their communities fit into efforts to combat global climate change. They demanded that the state address equity concerns in communities on the frontlines of climate change risks and impacts. State regulators responded with a market-based system that activists perceived as promising geographically neutral emissions reductions that could eventually trickle down to local communities. It contained no immediate mandate for direct emissions reductions in disadvantaged communities.

Trickle-down environmental policy was purported to be at its strongest in the constitution and decision-making of the California Economic and Allocation Advisory Committee, a body established by Governor Arnold Schwarzenegger's administration in 2009. The 16-member panel was composed of economists and climate experts charged with providing guidance to the California Air Resources Board in measuring the economic impacts of the AB 32 scoping plan and the best methods to implement a cap-and-trade program. At the economic committee meetings, environmental and social justice groups such as the Ella Baker Center, the Asian Pacific Environmental Network, and the Center on Race, Poverty, and the Environment lobbied the committee to endorse the creation of a climate fund. But during their deliberations, some committee members dismissed the idea as unnecessary and potentially costly.[12]

Environmental justice groups had a long road ahead to persuade policymakers to consider new forms of institutional design for climate action. One major hurdle was the disparate way in which values such as fairness were defined by environmental justice groups and administrative bodies that were often dominated by business interests, climate scientists, and traditional environmental groups. While advocates focused on alleviating the local harm caused by pollution, business groups defined fairness in terms of the lowest economic burden for polluters. When the board approved the AB 32 scoping plan, it chose not to adopt any measures to compensate environmental justice communities for the potential impacts of a cap-and-trade program. The board asserted that the most appropriate use for revenue generated under AB 32 was to retain it within or return it to the industry sector from which it was generated. The agency concluded that cap-and-trade was consistent with its environmental justice policies and would equally benefit residents of any race, culture, or income level.[13]

Anticipating that the board would not develop an equitable mechanism to distribute cap-and-trade revenue, environmental justice advocates pursued a parallel approach during the AB 32 implementation process. In a strategic move, they chose to return to their Latina/o allies in the legislature to ensure that the carbon market established by the state would have a statutorily mandated climate change community benefits fund. This strategy was first articulated by Shankar Prasad, a science fellow for the nonprofit advocacy group Coalition for Clean Air. Prasad had previously served as deputy secretary for science and environmental justice at the California Environmental Protection Agency. As a scientist and physician, Prasad had more than 20 years of environmental health and public policy experience. During his time with the agency, Prasad had witnessed the passage of dozens of environmental justice–focused bills that were neutered during the regulatory implementation process because they lacked a legislative mandate or clear funding mechanism to support environmental justice goals.[14]

Fearing that the environmental justice community would face a similar fate with cap-and-trade revenue, Prasad cowrote a concept paper. Titled "AB 32 Community Benefits Fund to Reduce Cumulative and Disproportionate Impacts," it argued that the state should be required to invest revenue generated by any carbon pricing mechanism (whether a carbon tax or cap-and-trade) in climate solutions that simultaneously tackled unhealthy local air quality and global greenhouse gases. Armed with this proposal, he approached environmental and social justice groups, such as the Ella Baker Center, the California chapter of the National Association for the Advancement of Colored People, and the Center on Race, Poverty, and the Environment, about sponsoring legislation to create a fund.[15]

Though these groups were apprehensive about returning to the legislature, they understood that real change would require a legislative mandate. According to both a senior capitol staffer and a Schwarzenegger appointee, advancing the climate fund legislation was a delicate issue for environmental justice groups, many of which were skeptical about cap-and-trade. Some, including the Center on Race, Poverty, and the Environment and other members of the Environmental Justice Advisory Committee, were preparing a lawsuit against the board in 2009 that aimed at forcing it to study alternatives to cap-and-trade and take into consideration the public health of disadvantaged communities (see Chapter 3). The capitol staffer later described the dilemma: "I think they were concerned that by engaging in this [legislation] that they might be validating the cap-and-trade program. And . . . that if we were moving forward with legislation, we might also be undermining their efforts at the time." Despite these complications, the Coalition for Clean Air chose to press forward with the legislation. In another capitol staff member's view, "A carbon tax was unlikely, and it didn't get the traction that cap-and-trade was getting. So if [California was] moving forward with cap-and-trade, we should at least guarantee that we are making the commitment in AB 32 for direct investments in disadvantaged communities."[16]

The Center on Race, Poverty, and the Environment, along with the National Association for the Advancement of Colored People, the Ella Baker Center, and the Greenlining Institute (a social justice nonprofit advocacy group), eventually agreed to cosponsor the proposal with the Coalition for Clean Air. In January 2009, they returned to the capitol to approach a legislator about drafting the climate fund legislation. With Assembly Speaker Fabian Núñez (author of AB 32) no longer in public office, environmental justice groups turned to Núñez's childhood friend, Assembly Member Kevin de León, a former union organizer and immigrant rights advocate. The cosponsors trusted de León because "he represents a community that obviously would be impacted by climate change, . . . and he had already demonstrated some leadership on the issue."[17] Throughout his legislative career, de León had made a commitment to improving air quality in his Northeast Los Angeles district, which is crisscrossed by six freeways and has some of the worst air quality in the nation.

In an attempt to secure his childhood friend's legacy and fulfill the promise of AB 32, de León agreed to draft the legislation, AB 1405. During the introduction of AB 1405 in February 2009, he emphasized that revenues set aside in the climate fund would give access to health and environmental cleanup funds to the state's most economically disadvantaged and polluted communities. He argued that California was "looking at a win-win for these communities—cleaner neighborhoods and better jobs with skills that will be in demand in the new economy."[18]

Opposition toward cap-and-trade shaped the terms in which the legislation was framed. According to Sofia Parino, the Center on Race, Poverty, and the Environment's senior staff attorney, her organization ensured the climate fund legislation was written in a fiscally neutral manner that did not link it to cap-and-trade. The legislation was drafted to simply state that the climate fund would receive revenue from AB 32, which at the time could have included profits from a carbon tax, since the board had yet to adopt cap-and-trade regulation. The problem, as Parino saw it, was how to support the environmental justice aims of the legislation without legitimizing the board's preference for cap-and-trade: "This is while we were still working on the AB 32 litigation and thought there was a possibility that we could get something other than cap-and-trade. We saw some of the groups that were bringing forward this legislation were not environmental justice groups in the beginning, . . . so we felt that we needed a voice and somebody that had a connection to the community. . . . We were really fighting that we didn't want [AB 32] tied to cap-and-trade."[19]

Advocates heralded the introduction of AB 1405 as the first attempt to fulfill the promise of AB 32 to "direct public and private investment toward the most disadvantaged communities in California."[20] For advocates such as the Asian Pacific Environmental Network's Mari Rose Taruc, sponsoring AB 1405 was a significant step toward building institutional capacity at the state capitol to bring back funds to local communities. Taruc's organization hoped to prove, through engagement and experimentation, the value of committing significant resources to developing a multiscalar advocacy campaign. AB 1405's cosponsors claimed two major areas in which it established an important precedent. The first was a minimum threshold of 30 percent of revenues generated under AB 32 to be deposited in a climate fund for investment in disadvantaged communities to accelerate greenhouse gas emissions reductions, mitigate co-pollutants, and create jobs. The second was a requirement that the board adopt a scientific methodology to identify the state's poorest and most polluted communities.[21]

However, as AB 1405 moved through the legislative process, it encountered strong opposition from polluting industries. In the bill's first assembly policy committee hearing in April 2009, the California Manufacturers and Technology Association dismissed the legislation as premature and unsubstantiated. Attempting to safeguard their interests, its members argued that the board should conduct additional studies before taking legislative action: "[The board] must balance cost-effectiveness, co-pollutant impacts, and technological feasibility as they develop the regulation. AB 1405 allows [the board] to ignore these criteria, and there is no economic or environmental analysis to justify the bill. An arbitrary amount of funding for broad unrelated purposes with unknown economic and environmental impacts is the bill's shaky foundation."[22]

In a good-faith effort to address these concerns, the cosponsors of the bill met with the California Manufacturers and Technology Association, the California Chamber of Commerce, and the Western States Petroleum Association. In the meeting, industry representatives expressed their support "in concept" for a climate fund but argued that it could siphon off billions of dollars from cap-and-trade revenue. When advocates pressed industry representatives on what they felt would be an appropriate amount, the representatives suggested that "a couple $100,000 or so could be placed in the pot." To environmental justice advocates, it became apparent that the polluting industries had a sense of ownership over the potential AB 32 revenue. As an environmental justice advocate aptly put it, "I think some of the industries felt like they were the ones paying the fees, and it was their money. They felt that they shouldn't have to give it to anybody, and industry should have control over it."[23] In the view of AB 1405's supporters, industries were paying for the right to pollute. Because industries were polluting a public good—the environment—the revenue generated under AB 32 was public and should be invested in ways that served the public interest.

These fundamental differences precluded compromise, and the California Manufacturers and Technology Association and the California Chamber of Commerce continued to oppose the bill. To counter industry opposition, the environmental justice groups sought to expand their base of support. They focused on developing a coalition of diverse advocacy organizations addressing public health, immigrants' rights, transportation, and affordable housing. They garnered these groups' support by making the case that community-based climate solutions could benefit their respective constituencies. The resulting coalition helped ensure that the assembly passed AB 1405 and moved it to the California State Senate. By the time the bill reached the senate floor, AB 1405 had 16 legislative coauthors and support from more than 30 diverse organizations. In response, the business lobby increased its opposition. Action alerts were sent out to the business community, urging it to contact key legislators, and prominent industry lobbyists were seen actively working senators to vote against the bill. The California Manufacturers and Technology Association accused Assembly Member de León of working on a "money grab" before the board had even implemented a cap-and-trade system.[24]

The business lobby's rhetoric that AB 32 should not be used to create a so-called "climate slush fund" hit home with many fiscally moderate senators. With less than a week left in the 2009 legislative session, AB 1405 lacked the votes to make it off the senate floor and to the governor's desk. According to Bonnie Holmes-Gen, policy director for the American Lung Association of California, this "very serious industry opposition to AB 1405"—especially the California Chamber of Commerce's condemnation of the bill as a "job killer"—had a significant impact.[25] The board's concerns about the legislation further diminished its prospects. The board argued that the legislation was premature since the Economic Allocation Advisory Committee had not developed its final recommendations. In a letter to de León dated August 13, 2009, the board stated that this committee had explicitly included equity as one of the issues to be addressed in its deliberations. They did not believe it was necessary for the legislature to determine the method or proportion of revenues to be allocated to any specific fund, at a minimum, until the committee had provided its recommendations.[26] This combined opposition led de León to withhold action on the bill until the following year. The sponsors felt that this delay would provide ample opportunity to convince fiscally moderate senators and the governor of the bill's merits.

Establishing Coalitions and Confronting Vetoes

While de León and the sponsors regrouped to determine how to proceed during the upcoming legislative session, the bill lost one of its main environmental justice sponsors, the Center on Race, Poverty, and the Environment. The center was the main litigant against the state's AB 32 scoping plan and found it difficult to justify support for legislation that most likely would be linked to a cap-and-trade system. By the fall of 2009, the center withdrew its support of AB 1405 and informed de León and other cosponsors that the move was intended to eliminate any perception of inconsistencies in its position on cap-and-trade. According to Sofia Parino, the center's support for AB 1405 could be seen as incentivizing and validating cap-and-trade: "I think about the same time when it moved to a two-year bill, it was also clear that cap-and-trade was going to be the mechanism. . . . So we decided we weren't going to sponsor the next iteration, because fiscally [AB 1405] was neutral, but in practicality it was [funded] from cap-and-trade. Our communities gave us a strong sense that they didn't want to have anything to do with it, that it was dirty money, . . . and one of our clients often stated that they didn't want a 'gold-plated inhaler.'"[27]

By taking the politically nuanced position of "neutral," the center could maintain its opposition to a cap-and-trade system without undermining the establishment of a Climate Change Community Benefits Fund. Perhaps in an attempt to maintain its allies and not burn political bridges in Sacramento, the Center on Race, Poverty, and the Environment helped secure the California Environmental Justice Alliance as an official cosponsor of AB 1405. California's six largest environmental justice organizations had just formed this statewide coalition to advocate at the state and local levels for policies protecting public health and the environment. As one of its founding members, the center recruited the coalition to ensure that AB 1405 supporters maintained a balance of expertise in policy development and on-the-ground experience. With the alliance anchoring an environmental justice perspective, additional environmental and social justice groups throughout the state joined as cosponsors.[28]

With the addition of new bill sponsors and de León settling into his leadership role as a more seasoned chairman of the powerful Assembly Appropriations Committee, the prospects of moving AB 1405 to the governor's desk for signature seemed brighter in the 2010 legislative session. De León spent the next several months lobbying key senators to gain their support for the climate fund bill. His growing reputation as an effective chairman in the assembly persuaded several senators from politically moderate districts to reconsider their position on AB 1405.[29] Moreover, in January 2010, de León announced his intention to run for the open 22nd state senate district seat in Los Angeles. The lack of major rivals portended victory. De León's rising political stature and the possibility of election to the state senate prompted some senators to develop a more collegial relationship with de León by supporting AB 1405. According to a senior capitol staffer: "His leadership was huge. He would walk the halls to go lobby members. He drove out to the Imperial Valley to go meet with a member. He offered to drive out to the Inland Empire to meet with a member. . . . He offered to drive up to Santa

Ana to meet with another member. He really worked it. He made several efforts to meet with the [governor's] administration. . . . We couldn't have picked a better author [for the bill]."[30]

After several months of lobbying, de León moved forward the climate fund legislation on August 19, 2010. After a subsequent amendment to reduce the required minimum investment threshold of 30 percent to just 10 percent, the author and cosponsors hoped the bill might be more palatable to fiscally moderate Democratic senators. When the amended version of AB 1405 came up for a floor vote on August 30, the new amendments and lobbying strategy paid off. AB 1405 passed the senate with 22 votes (one more than needed), almost on a strict party-line vote, with all Republicans and one fiscally moderate Latino Democratic senator from Orange County opposing it. AB 1405 returned to the assembly, which immediately approved it and sent it to Governor Schwarzenegger's desk for consideration.[31]

Despite this swift progress, AB 1405 did not escape controversy. During the assembly vote, Republican assembly members took the opportunity to denounce the bill and California's climate change program. In doing so, they attempted to bring the whole premise of environmental justice measures into question. They asked whether environmental justice communities were truly harmed by greenhouse gas emissions and why such communities deserved financial compensation from the cap-and-trade program. They further argued that the cap-and-trade program would actually hurt low-income communities of color by raising energy and fuel prices.[32]

The most vocal in making these objections was Republican assembly member Chuck DeVore, who declared that AB 1405 "lays bare what all this cap-and-trade is about regarding climate change . . . because it's not about the climate. At the end of the day, it's about power and taking money from certain industries and carving that money out and shipping it to politically favored groups."[33] In defending the measure, de León and other Democratic members countered that that the funding from AB 1405 would help compensate California's most environmentally burdened communities by reducing both greenhouse gas emissions and their co-pollutants. They further maintained that a climate fund could help create new green-collar jobs in these disadvantaged communities. Through this linkage of greenhouse gases, co-pollutants, jobs, and public health, the most vulnerable neighborhoods would not be left behind in the state's climate programs.[34]

While not as outspoken as assembly Republicans, the board continued to question the need for AB 1405. The board was still drafting regulations to implement a cap-and-trade system, which was on track for a November 2010 adoption, contingent on any potential court delays from the environmental justice lawsuit. Consequently, when the bill reached his desk, Governor Schwarzenegger echoed the board's position and vetoed the measure. In his veto message on September 30, 2010, he stated that the measure was premature and questioned the need for legislation to address the effects of climate change in disadvantaged communities. Schwarzenegger encouraged environmental justice advocates to work with the board during the pending regulatory process instead of sponsoring legislation.[35] According to a senior appointee in the Schwarzenegger administration, the governor vetoed AB 1405 because of the dynamics at play among the Environmental Justice Advisory Committee, the board, and the environmental justice litigants. There was a sense from the board leadership that if environmental justice groups were suing them, "why should we reward them?" The situation had become so politically charged by June 2010 that more than 100 environmental justice advocates protested outside the home of the board chair, Mary Nichols, over AB 32 implementation and the regulation of toxic diesel pollution from railway operations. Another factor in the governor's veto, according to the Schwarzenegger appointee, was lingering resentment toward environmental justice groups that had never come on as official supporters of AB 32 or joined the governor during the bill's signing ceremony in Malibu.[36]

The Scientific Instruments of Environmental Justice

I do have what I believe to be a major responsibility to protect those that have little or no voice. Whether they are being choked by freeways or stationary emitters of CO_2 or traditional pollutants, if they don't have a voice here [in the legislature],

> then they have no voice anywhere else. So I'm looking forward to a very close and proactive working relationship with you, Mr. Rodriquez.
>
> —SENATOR DE LEÓN to the California Environmental Protection Agency secretary, Matt Rodriquez, during his Senate Rules Committee confirmation hearing

Undaunted by Schwarzenegger's veto, de León, now a newly elected senator, was more determined to ensure that AB 32 revenue was invested in California's most disadvantaged communities. With the inauguration of Jerry Brown, a progressive Democrat, to his third term as governor in 2011 (Brown had served two terms from 1975 to 1983), de León believed the climate fund legislation had a much stronger possibility of being enacted in the upcoming legislative session. In moving new legislation forward, he wanted to get past the lawsuits and the collective grudges between board officials and environmental justice groups. De León also wanted to change the political dynamic, since he didn't see himself as part of either camp. According to a senior senate aide, de León understood his responsibility in representing communities that had been impacted by climate change and air pollution. By enacting climate fund legislation, he hoped to ensure that his communities were accounted for in the AB 32 cap-and-trade equation.[37]

On February 17, 2011, Senator de León reintroduced the climate fund legislation as Senate Bill (SB) 535. The legislation maintained the same cosponsors as the final version of AB 1405; however, it added an important political ally: the National Resources Defense Council. In the previous two years, this and other traditional environmental groups had supported the climate fund legislation but did not actively lobby for it or otherwise seek its passage. The addition of the council as an active cosponsor marked a turning point in gaining broader support. As an original sponsor of AB 32 and a staunch defender of a cap-and-trade system, the council provided additional capabilities and legitimacy for establishing a climate fund. In courting the council as a cosponsor, Senator de León also hoped to influence board officials and Governor Brown. In supporting

SB 535, the council took steps to make itself more relevant to the changing demographics in California and within the legislature (see Chapter 3).[38]

The new climate fund coalition helped SB 535 sail through the senate and assembly policy committees with broad support among environmental and community-based organizations. In voting in favor of SB 535, several lawmakers argued that the legislature should exert more direct control over key policies in the cap-and-trade program, rather than letting the board determine all the decisions. The Legislative Analyst's Office, the legislature's nonpartisan fiscal and policy advisor, had endorsed this reasoning in a report and at a senate budget subcommittee hearing. The office's Mark Newtown recommended that lawmakers take a more substantive role in establishing rules for the cap-and-trade program. He stated that the use and design of cap-and-trade mechanisms was complex and involved a lot of policy choices, which should be signed off by the legislature.[39]

Business and industry groups, on the other hand, continued to attack the legislation, arguing that it was jumping ahead of the board's regulatory proceedings. At an Assembly Natural Resources Committee hearing on June 27, 2011, California Chamber of Commerce lobbyist Brenda Coleman further questioned whether disadvantaged communities would suffer any additional damage to public health from implementation of a cap-and-trade program. She cited a recent board report that determined that the program would not cause an increase in co-pollutants. The California Council for Environmental and Economic Balance, an industry association, also asserted that SB 535 conflated the impacts from criteria air pollutants, such as particulate matter and ozone, and the alleged impacts from greenhouse gas emissions.[40]

Despite these attacks by industry, the legislature appeared poised to pass the climate fund legislation and send it to the governor in record time. Yet SB 535 met a political roadblock in the Assembly Appropriations Committee. On August 25, 2011, the committee held the legislation during a hearing on all pending fiscal bills. It was determined that the bill would be too costly for the state to implement. The move made SB 535 a "two-year" bill, ineligible for committee consideration until the next legislative session. With a second setback for the climate community benefits fund legislation, de León and the SB 535 cosponsors retreated and took the following year to determine their course of action.[41]

After several strategy meetings, the coalition decided to focus on building up their practical authority through the implementation of a key provision of the climate fund legislation. This involved working through the regulatory process at the California Environmental Protection Agency while awaiting the bill's fate in the Assembly Appropriations Committee. The various iterations of the climate fund legislation had required the state to adopt a scientific methodology for identifying and investing in environmental justice communities. Advocates envisioned developing this methodology from the efforts they had made over nearly a decade to commit California to a comprehensive policy on cumulative impacts (see Chapter 2). One central element would be a tool that would systematically and spatially identify communities most burdened by multiple sources of pollution and vulnerable to its effects. Advocates felt that by identifying such communities across California, the state would better understand where to target limited resources and programs.

Since 2003, statewide policies highlighting the significance of cumulative impacts had been on the books. Yet political leaders in Sacramento avoided adopting a state-sanctioned cumulative impacts tool. In an attempt to break through this impasse, the California Environmental Justice Alliance formed a strategic partnership in 2010 with the academic researchers Manuel Pastor (University of Southern California), Rachel Morello-Frosch (University of California, Berkeley), and Jim Sadd (Occidental College) on the ongoing development of their cumulative impacts tool, the Environmental Justice Screening Methodology. For the past several decades, these individuals have been considered the leading environmental health scholars in the country and have developed new methods to quantify the conditions that most concern environmental justice advocates. The board funded the initial analytical work on the screening methodology and supported it through input from agency scientists and an external scientific peer review committee. But the team of scholars had also consulted environmental justice groups regarding their interpretation of preliminary results and the selection of appropriate indicators. This strategy—soliciting peer review from agency personnel, scientific experts, and community stakeholders—sought to ensure that the final screening methodology was scientifically sound and transparent to diverse regulatory, policy, and advocacy audiences.[42]

The California Environmental Justice Alliance also contracted with the same researchers in 2011 to field-test the draft screening methodology in 11 geographically diverse environmental justice communities throughout California. In the process, they organized more than 70 community residents and 30 community organizers to "ground-truth" the model and test its validity. The final result was developed as a collaborative, community-academic research model, with residents and advocates in each of the targeted geographic areas trained to verify the accuracy of data and incorporate local knowledge into the model. This approach, while intended to verify the results, also could help inform community members about their environmental surroundings by building local capacity to understand the data and use it for policy advocacy. It also provided on-the-ground observations from residents that public databases often overlook. For example, the board had several definitions for "sensitive sites" and "hazardous receptors," but the California Environmental Justice Alliance pointed out that those definitions might overlook places that residents identified as hazardous or sensitive, such as nail salons and dry cleaners. The coalition considered such ground-truthing as a method to check the accuracy of "official" government data, which could contain erroneous facility locations or omit them altogether.[43]

The environmental justice advocates described the screening methodology as including the most comprehensive indicators of environmental, health, and socioeconomic factors to develop a relative scoring system to assess cumulative impacts at the neighborhood level. Its scoring system used 23 indicators, which had been identified through scientific research and community-based input and organized into three categories: hazard proximity and land use, air pollution exposure and estimated health risk, and social and health vulnerability. Using geographic information system mapping techniques and land-use data, the screening methodology assesses the number and volume of indicators present and their proximity to sensitive land uses in a particular census tract. This screening is used to develop a cumulative impacts score and color-coded maps to visually depict the intensity of cumulative impact. The field tests covered more than 15 square miles of California environmental justice communities, in both urban and rural settings.[44]

The Center on Race, Poverty, and the Environment and professor Jonathan London of the University of California, Davis, supplemented this work through the development of the Cumulative Environmental Vulnerability Assessment, which was specific to environmental concerns in the rural San Joaquin Valley. This dual engagement allowed environmental justice groups to compare the two models, understand how the tools were best applied, and increase their comprehension of the science of cumulative impacts. After years invested in developing and testing methodologies, the California Environmental Justice Alliance and its allies considered themselves experts on cumulative impacts and understood how the state could implement a science-based tool. Their screening methodology field work indicated that low-income neighborhoods and communities of color across California are subject to high poverty rates, surrounded by hazardous land uses, and exposed on a daily basis to emissions from toxic industries. This data confirmed, with greater scientific credibility, what activists had argued all along.[45]

Environmental justice advocates saw the screening methodology maps as a scientific and political means of communicating to both local people and regulators about the environmental health conditions in their communities. According to the urban planning scholar Jason Corburn, maps strongly shape public decision-making. Maps are not objective, but they often become the prevailing image of reality by resonating with those in political power. They also perform at least three political functions in relation to knowledge production and policy implementation. First, maps "aggregate and select data," and different methods result in different interpretive outcomes. The screening methodology maps offered a way to visualize advocates' concerns, which were obscured in more conventional assessments because these often focused only on individual chemicals and their sources. Second, maps are "identity forming tools," since their symbols are used to visually present data and render particular representations. Through the screening methodology, the ranking of neighborhoods within regions helped define local areas that might need targeted regulatory strategies to address environmental justice concerns. Third, maps serve as "boundary makers" through the inclusion or omission of certain data. By creating boundaries around "what is and is not important to see, maps can encourage viewers to 'see like a state' or suggest some other imagined vision."[46] In this capacity, according to an article by Sadd, Pastor, and Morello-Frosch, the screening methodology can help relieve communities of the burden of proving that cumulative environmental impacts exist. The maps provide direct and detailed evidence to decision makers, so that residents without an active environmental justice movement or capacity for civic engagement can also benefit from regulatory relief.[47]

Moreover, the screening methodology served as a political tool that provided increased capabilities and recognition for environmental justice groups by pressuring state regulators at the California Office of Environmental Health Hazard Assessment and the board to think about how disadvantaged communities were impacted, as well as methods to define and locate those communities. In response, regulators developed their own draft cumulative impacts tool, partly based on the screening methodology. However, they chose to bypass a community-based participatory research approach. State scientists truncated several indicators in the screening methodology tool (such as race/ethnicity and climate vulnerability metrics) and focused on a regional screening approach based on zip codes, instead of a neighborhood-scale assessment using census tracts. According to a senior agency appointee, they did not include race/ethnicity as an indicator because it was anticipated that the tool could be used to distribute state grants. The appointee cited the California Civil Rights Initiative (Proposition 209), approved by voters in 1996, which prohibits state agencies from using racial/ethnic preferences in governmental programs and decisions. In releasing their draft cumulative impacts tool, regulators indicated that it would need more input and analysis from a broad set of stakeholders, including representatives of industry, before state agencies adopted a final version.[48]

The realization that this process could take several years angered the California Environmental Justice Alliance and its allies. They believed that the state had no reason to delay the implementation of a cumulative impacts tool, since the best available science on cumulative impact screening already existed. In a comment letter to regulators, the alliance argued that the agency's adoption of zip codes as a standard of scale would obscure the worst-affected communities. Since zip codes cover large geographic areas encompassing neighborhoods with vastly different levels of pollution and environmental degradation, environmental justice communities were in danger of being "washed out." The alliance identified 11 other key deficiencies in the state's cumulative impacts tool, including the lack of a pesticide exposure indicator. Moreover, it asserted that the screening methodology was ready to implement, based on peer review and thorough field testing. It claimed that regulators would be engaging in a poor use of limited state resources if they began a time-consuming regulatory process to finalize yet another tool: "California Environmental Justice Alliance supports cumulative impact tools in statewide policy; the time has come to take action on this critical issue. For years, a concern with incorporating cumulative impacts into policy-making has been that we lack the scientific basis to identify communities that face disproportionate impacts. We now have the science."[49]

According to a former senior capitol staff member, state regulators may have stalled the implementation of the cumulative impacts tool because of opposition from industry. It was perceived that Pastor, Sadd, and Morello-Frosch, along with the California Environmental Justice Alliance, developed the screening methodology without significant input from the private sector, and as such, businesses challenged its scientific validity. Fearing that the mapping tool would result in decreased real estate sales or possible legal liability, industry groups argued that it was biased and arbitrary and would potentially distort findings. For example, some industry representatives worried that the screening methodology placed too much weight on measures that might not reflect actual exposures, such as proximity to hazardous waste sites.[50] Industry groups for years have expressed opposition to the use of cumulative impacts studies, with the argument that such studies could prompt additional environmental reviews and health assessments beyond the scope of existing regulations (see Chapter 2). At a public meeting held to review the state's draft cumulative impacts tool, a representative from the Chemical Industry Council of California alleged that "the obvious intent of some is to push [cumulative impacts tools] into the regulatory realm." Other business representatives at the meeting warned that communities with high impact scores might scare away industry and economic investment.[51]

According to Mari Rose Taruc, the California Environmental Justice Alliance understood the implications of this growing opposition. In a pragmatic move, it shifted its focus to support regulators in finalizing the state-developed cumulative impacts tool. They amended SB 535 to provide the California Environmental Protection Agency with only broad criteria to identify disadvantaged communities, which included (1) areas disproportionately affected by pollution and other hazards that can lead to negative public health effects, exposure, or environmental degradation; and (2) areas with a concentration of low-income people who suffer from high unemployment rates, low homeownership levels, high rent burdens, or low educational attainment levels.[52] By proposing flexibility in the state's development of a cumulative impacts tool, advocates anticipated a smoother route to passage for SB 535. As Taruc later recounted, adopting any such tool would itself be a major step forward: "We thought that the [screening methodology] was a superior tool, . . . but we understood that identifying cumulative impacts in communities across the state was big. . . . We get asked by policymakers, what is environmental justice, and where are these areas? For us to be able to point to a scientific tool that lifts up what the community knows but isn't always supported by research was just great to have."[53]

The upcoming Senate Rules Committee confirmation hearings on the selection of Matt Rodriquez as secretary of the California Environmental Protection Agency, George Alexeeff as director of the California Office of Environmental Health Hazard Assessment, and several board appointees formed part of another key environmental justice strategy. The Senate Rules Committee, of which de León was a member, holds confirmation hearings on gubernatorial appointees in fulfillment of the senate's constitutional "advice and consent" responsibilities. Although the vast majority of confirmation hearings are routine, nominees can be rejected if they are viewed as out of alignment with the senate's policy priorities and values. Through the senate confirmation process, SB 535 supporters hoped de León could get a commitment from Rodriquez to implement a cumulative impacts tool within the next year. Moreover, such a commitment could also bolster the probability that SB 535 would get Governor Brown's signature, as the legislation would be consistent with his administration's ongoing regulatory efforts.[54]

During the confirmation hearings and in private meetings, de León questioned nominees on their willingness to implement a cumulative impacts tool and their support for investing cap-and-trade funds in disadvantaged communities. As part of this line of questioning, de León placed gubernatorial appointees on the public record regarding their positions. The strategy proved effective. In his confirmation hearing on February 12, 2012, Rodriquez expressed his support for a cumulative impacts tool and committed to seeking compromise with the legislature on investing cap-and-trade revenues in disadvantaged communities: "One of the things I want to work on, not just in AB 32 but generally, is to continue to get information so that we can identify disadvantaged communities in the state. Why do we need this information? So that we can then focus programs of various kinds to help areas that are disadvantaged. . . . I have been unable to come to a position on what floor or threshold we should have [in regards to investing cap-and-trade funds in impacted communities]. . . . I know we are going to have to come to an agreement with the legislature on how these funds can be expended."[55] According to a senior capitol staff member, the commitment that de León extracted from Secretary Rodriquez and other gubernatorial appointees was instrumental in getting regulators to agree to adopt a cumulative impacts tool. Staff scientists and Brown administration appointees, who were previously reluctant about this requirement, were now more likely to follow through.[56]

Enacting and Implementing the Climate Community Benefits Fund

Several more months would pass before SB 535 could move forward. When the Assembly Appropriations Committee released the bill on August 16, 2012, it was on the condition that de León accept substantive amendments. The new amendments deleted the establishment of the climate fund and instead required depositing cap-and-trade revenue in the Greenhouse Gas Reductions Fund established by a companion bill, AB 1532.[57] The amendments mandated not only that 10 percent of cap-and-trade revenues be directly invested in projects located in disadvantaged communities but that at least 25 percent of revenues should be directed to projects providing general benefits to disadvantaged communities. Through AB 1532, the board and the state Department of Finance were also directed to develop an investment plan every three years for auction revenues, using a public process that would maximize economic, environmental, and public health benefits within California.[58]

By the time SB 535 and AB 1532 came up for their respective floor votes in the assembly and senate, both measures had the support of three former California Environmental Protection Agency secretaries and two former board chairs, as well as endorsements from nearly 200 public health, labor, clean technology and energy, environmental and social justice, and immigrants' rights organizations. Most notably, the California Latino Legislative Caucus identified both measures as "priority bills," and several ethnic minority chambers of commerce added their support. These changes also prompted several key fiscally moderate Latina/o legislators, who had previously opposed SB 535, to reverse course. The bills eventually passed through the assembly and senate, with Governor Brown signing them into law on September 30, 2012—one month after the Richmond oil refinery fire (see Chapter 2). Environmental and social justice groups hailed their signing as a major victory. It validated a climate change policy agenda that community advocates had spent years developing. As the Ella Baker Center announced in a blog post the day after Brown signed the bills into law, "Our work on SB 535 began in 2009 when our [Oakland] Green Collar Jobs Campaign saw a way to ensure that Cap and Trade revenue could actually make a difference in California communities most hurt by poverty and pollution. . . . [T]hanks to the hundreds of you who have written, called, and showed up in Sacramento to advocate for these vital bills. Together we have secured a brighter, greener future for California communities that need it the most."[59]

The celebration was short lived, however, as the cosponsors shifted attention toward implementation. Environmental justice advocates understood that passing the law was only the first step; the real work would begin with state agency implementation. This crucial phase included requirements that the board and the Department of Finance develop a three-year investment plan to allocate the billions of dollars raised by the cap-and-trade program. The agencies were scheduled to release a draft of the plan for public comment in early 2013. According to a key supporter of the bills, "After it was signed into law, the co-sponsors, [the Asian Pacific Environmental Network], the Coalition for Clean Air, Public Advocates, and the Greenlining Institute, then said that we want to continue with implementation. So we turned our supporter list from SB 535 into an implementation coalition—the SB 535 Coalition."[60]

The SB 535 Coalition convened strategy meetings in October 2012 with grassroots organizations about developing a framework to secure the benefits of SB 535 for disadvantaged communities. From those meetings, the coalition went on to conduct several webinars and regional and statewide workshops. These workshops helped facilitate engagement in the board's public process for adopting the investment plan; they also gathered ideas from disadvantaged communities about potential funding opportunities. Chief among these efforts was a November 2012 survey of 28 environmental justice organizations. The survey results helped the coalition better understand the needs of California's disadvantaged communities and which programs should be prioritized in the board investment plan. The survey results identified five existing statewide programs as high priorities for near-term investments: sustainable transportation, affordable housing, low-income energy efficiency programs, renewable energy programs, and urban forestry projects. The coalition chose these existing programs because they represented opportunities for investments that could provide the greatest health and economic benefits in disadvantaged communities.[61]

Throughout the winter of 2013, the SB 535 Coalition developed comprehensive reports, fact sheets, and lobbying strategies aimed at including their funding priorities in the investment plan (fig. 5.1). The strategy extended to organizing environmental justice and social justice leaders' appearances at the board's investment plan hearings and submitting comment letters that highlighted their recommendations. In lobbying for their priorities, the coalition did initially encounter resistance to their proposal to fund affordable housing near transit as a climate mitigation solution. According to an SB 535 Coalition member, administration officials doubted that affordable housing could play a meaningful role in climate action planning: "When we first started talking on the state level about affordable housing, it wasn't even on the radar of a lot of decision makers as a strategy. . . . I think people weren't seeing the links between greenhouse gas reductions and affordable housing and also the links between affordable transit and actually maintaining existing ridership."[62] Many policymakers saw the construction of housing projects near transportation as an effective greenhouse mitigation measure. Yet they did not understand that it also threatened to displace existing residents. Without the inclusion of affordable housing units, transit-oriented development projects could undercut greenhouse gas mitigation strategies by forcing lower-income people to relocate to cheaper suburban locales where they would be dependent on cars for transportation (see Chapter 4).[63] Preliminary research developed by the Oakland-based transportation justice nonprofit TransForm and the California Housing Partnership supported these conclusions. Their analysis showed that a 10 percent investment of cap-and-trade revenues could create 15,000 affordable housing units near transit. This equates to nearly 2 million metric tons of greenhouse gas reductions, or 105 million fewer miles driven over the estimated 55-year life of the buildings. Their analysis also showed that low-income households living within a quarter-mile of transit drove 50 percent fewer miles than higher-income households living in an equal radius. Through such studies, advocates argued that affordable housing projects could serve as a scientifically valid greenhouse gas reduction measure that also addressed equity concerns.[64]

After months of lobbying and presentations of studies linking housing displacement and increased greenhouse gas emissions, the SB 535 Coalition persuaded the board, the Department of Finance, and the governor's office that investing in affordable housing near transit was an effective climate mitigation strategy. In May 2013, the board released its first investment plan for cap-and-trade revenue for legislative review and approval. The plan provided guidelines for prioritizing investments that would achieve greenhouse gas reduction goals and yield valuable co-benefits at the statewide scale and, most importantly, in disadvantaged communities. Through a process

SB535 Coalition

5 Priority Programs for Near-term Investments

After Governor Brown signed SB 535 (de León) and AB 1532 (J. Perez) in 2012, the SB535 Coalition quickly began the work of engaging grassroots, community-based organizations and individual supporters across the state to better understand the needs of disadvantaged communities in response to causes and effects of climate change. These are our top 5 near-term program ideas that should be funded by the Greenhouse Gas Reduction Fund.

Community Greening

(ie CalFire Urban and Community Forestry Program) Continue the Urban and Community Forestry Program technical assistance and local assistance grants as a mechanism to increase tree-planting, education, and proper tree management.

Low-Income Energy Efficiency Programs

(ie Energy Savings Assistance Program, Weatherization Assistance Program) Reduce energy costs for low-income families, help overcome program limitations, expand the suite of efficiency measures and explore pilots to reach customer segments currently underserved by these programs, including low income tenants in multi-family housing.

Renewable Energy

(ie Single and Multi-Family Affordable Solar Homes program - SASH/MASH) Solar incentives for electric utility customers directed toward low-income and affordable housing programs to stimulate the adoption of solar power and decrease electricity usage and bills.

Transit Operations

(ie State Transit Assistance - STA) Increase transit ridership and reduce GHG emissions by targeting funds to operating increased levels of transit service and implementing fare reduction strategies that incentivize greater transit utilization.

Transit-Oriented Development

(ie Affordable TOD Housing program) Provide transit-oriented housing affordable to very-low income households, who own fewer cars and use transit at significantly higher rates than the general population.

The SB535 Coalition is led by the Asian Pacific Environmental Network, Coalition for Clean Air, Greenlining Institute & Public Advocates Inc. For more info, contact Mari Rose Taruc at marirose@apen4ej.org

Figure 5.1a. SB 535 priority funding fact sheet. Courtesy of the California Climate Equity Coalition (formerly the SB 535 Coalition)

SB535 Coalition

Recommendations for Maximizing Job Creation and Workforce Co-Benefits

AB 1532 (Perez) and SB 535 (de Leon) provide that AB 32 revenues should meet the following goals:

- Foster job creation by promoting in-state greenhouse gas emissions reduction projects carried out by California workers and businesses
- Direct investment toward the most disadvantaged communities and households in the state
- Maximize economic benefits for California wherever possible

Cap-and-trade revenues should meet the above goals by implementing the following recommendations:

Create Jobs for Disadvantaged Californians

Cap-and-trade funds should be subject to existing local and targeted hiring agreements. The Legislature should require implementing state agencies to adopt local/targeted hire policies in the use of these funds.

Include Skill Certifications Requirements on Use of Funds

Many of these jobs will be in the construction industry, especially in low-carbon transportation and energy sectors. Clean energy and Infrastructure projects should meet the highest standard of quality control to maximize GHG reductions and jobs benefits. Cap-and-trade investments, along with leveraged funds, should include explicit standards for contractors and minimum training and skill standards for workers. These requirements should be identified by the agencies with input from the key stakeholders.

Set Performance Goals, Track Data, and Require Program Evaluation

The Legislature should require that implementing agencies set explicit goals for the quantity and quality of jobs created and the demographic and geographic distribution of workers, particularly those in entry-level jobs. Performance metrics and job reporting requirements should be developed using a common cost-effective state-wide computerized job tracking and labor compliance system. Evaluation of the jobs and workforce development outcomes of publicly-funded programs is critical to insuring accountability and informing program performance. Evaluation should include quantity and quality of jobs created, including wages and benefits, and the demographic and geographic profile of workers.

Figure 5.1b. SB 535 priority funding fact sheet. Courtesy of the California Climate Equity Coalition (formerly the SB 535 Coalition)

of engagement, experimentation, and scaling up, environmental justice groups gained practical authority and recognition from regulators for many of their proposed climate solutions, thus transforming ideals once seen as impractical into feasible measures.[65]

In the first year of SB 535 implementation, California invested nearly $300 million for greenhouse gas mitigation projects benefiting disadvantaged communities. These projects included urban forestry, low-carbon transportation, transit operations and affordable housing, energy efficiency programs, and waste diversion. The SB 535 Coalition worked closely with regulators to clarify criteria and processes for determining which projects to consider under SB 535 and for the defining of co-benefits. Due in large part to the coalition's efforts, the board's guidance memo for agencies administering investments emphasized projects that spurred local green-collar jobs and improved public health—in particular by addressing asthma and obesity.[66]

With the adoption of SB 535 and the cap-and-trade investment plan, the concept of a climate fund had come full circle. Although California's cap-and-trade program still remains highly controversial with many environmental justice groups, the SB 535 Coalition deepened the state's commitment to disadvantaged communities by making environmental justice central to climate change policies and investments. Mari Rose Taruc later described how, with the passage of SB 535, the Asian Pacific Environmental Network's work on state-level climate policy "has to come back to Richmond and Oakland. . . . We have to be able to see those benefits come back to the communities we are working in." The climate fund provides community-driven opportunities to link California's climate change policies with efforts to improve air quality, public health, and local economies.[67] In contrast to the top-down approach initially taken in the AB 32 scoping plan, SB 535 represents an experiment that challenged institutional practices and created new, multipurpose solutions. Those collaborative solutions, in turn, may help California's climate change programs begin to address inequities at multiple scales.

Rescaling Climate Change Policy and Decision-Making

The adoption of California's Climate Change Community Benefits Fund illustrates how and why particular governments link global climate change with equity and local health. A crucial component of those linkages was the capacity of environmental justice groups to develop their practical authority. Advocates such as the Asian Pacific Environmental Network's Mari Rose Taruc and Emily Kirsch of the Ella Baker Center convinced decision makers about the ways in which local experiments could contribute to the learning and sharing of knowledge across levels and scales. Through experimentation, environmental justice groups not only contributed to the building of their practical authority but also changed regulatory culture. The passage of the climate fund legislation and the strategic implementation lobbying efforts pressured state regulators to engage with advocates to validate community-based solutions. In the process, environmental justice groups, scientists, and policymakers creatively adapted institutional rules, such as those of AB 32, that had originally been made for other purposes (for example, global greenhouse gas emissions reduction) in their attempts to resolve seemingly intractable environmental problems at multiple levels.[68]

The adoption of the cumulative impacts tool, in particular, placed a scientific inscription of environmental justice within California's climate change governance. It enhanced environmental justice groups' practical authority by providing them with a range of knowledge production processes, such as government statistics, quantitative data, and mapmaking, to use for public reasoning and decisions around environmental issues. According to Sasha Barab et al., by inscribing environmental justice values within a scientific methodology, the tool affects the manner in which people and institutions conceptualize the dynamics of environmental problems. Simplified representations such as the cumulative impact maps thus are essential elements in translating science into policy action and decisions about the larger context from which these values arose. Thus, the cumulative impacts tool is a conceptual instrument for making sense of the world.[69] Its use in climate change governance has the potential to shape regulators' technical practices and further ensure that climate solutions not only have a global benefit but also—just as importantly—bring positive local outcomes to the most disadvantaged communities.

California advocates have made significant progress in altering the state's regulatory culture of climate change. Initially they attempted to do so through oppositional tactics, including the filing of lawsuits. The case of the Climate Change Community Benefits Fund sheds light on how institutions and decision-making structures can be transformed to facilitate collaborative outcomes. As the prominent environmental justice scholar Robert Bullard points out, "The impetus for changing the dominant environmental protection paradigm did not come from within regulatory agencies, the polluting industry, academia, or the 'industry' that has been built around risk management. The environmental justice movement is led by . . . leaders who question the foundation of the current environmental protection." However, at the same time, advocates working within regulatory processes also often find that they have to change their stance by engaging in a series of "trade-offs" to bring about social transformation.[70]

Institution-building around climate change is an exceedingly messy process. It requires activists, policymakers, regulators, and polluting industries to reach compromises and develop shared practices. Equitable forms of climate change governance require a willingness to explore new ideas about how to protect nature and local communities. While most environmental justice groups are philosophically opposed to market-based mechanisms, social movements such as the SB 535 Coalition embrace pragmatic approaches to ensure that the programs they support are equitably funded and implemented. The coalition's decision to develop a community benefits fund, rather than reject it on principle, was crucial to this type of practical innovation. Climate change governance, thus, is not a value-neutral exercise but instead involves political processes embedded in cultural contexts.[71]

The implementation of SB 535 compared to the contentious fate of the AB 32 scoping plan shows that climate change policy can only achieve public credibility and authority when the boundaries of the relevant moral-political spaces are redrawn so as to accommodate the interests of multiple actors. Various forms of experimentation enabled environmental justice groups to engage collaboratively with regulators and legislators to define local expressions of the state's climate change program. This process would continue, with alliances forged across national boundaries, as California policymakers sought to create a far-reaching carbon market in the Global South.[72]

CHAPTER SIX

Climate beyond Borders

IN THE MOVIE *Predator*, Arnold Schwarzenegger plays Major Dutch Schaefer, a soldier of fortune hired by the US government to rescue a group of diplomats held hostage in a Latin American jungle. But things go wrong for Dutch and his military dream team when they arrive to carry out their mission. They find dozens of dead bodies and discover they are being hunted by a ruthless creature with supernatural strength. Here, nature itself is the threat—the predator's camouflage makes it feel like the "good guys" will be consumed by the jungle at any moment. Eventually, Dutch's entire team is killed; he defeats the alien predator and saves the world. Filmed in the tropical forests of Chiapas, Mexico, *Predator* grossed nearly $100 million worldwide and established Schwarzenegger as the quintessential action hero. He went on to parlay that success to his ultimate role. In 2003, Schwarzenegger replaced then-governor Gray Davis in a special recall election. He won reelection as California's governor in 2006 in part for signing into law the Global Warming Solutions Act (AB 32). The measure created the nation's first cap on greenhouse gas emissions and provided Schwarzenegger with global acclaim as a fierce climate action hero.

A few years later, I would witness the movie star turned climate action governor up close at the Beverly Hills Hilton. In 2008, Schwarzenegger hosted the Governors' Global Climate Summit, which brought together a dream team of US and international governors and diplomats. Hundreds of attendees from more than 50 states, provinces, and countries joined the two-day meeting, where participants signed a declaration outlining their plans for turning climate goals into action. Schwarzenegger took to a shimmering stage where he described climate change in terms akin to his on-screen nemesis in *Predator*: an existential threat to humanity. Time and again, he framed himself and California as the ultimate climate warriors in protecting the world. During a coffee break, I got up to walk around the Hilton's International Ballroom. I realized it was one of Hollywood's most famous event spaces: the site of the Oscar Nominees Luncheon and the annual Golden Globe Awards.

During the climate summit, I heard an international chorus of famous voices billing the event as a grandiose path toward "thinking as one planet," which fit with the ballroom's pedigree. Governor Schwarzenegger would tout his global mentality that allowed him to relate to countries in the Global South that have different challenges, struggles, histories, and economies.[1] He stressed the need to combat tropical deforestation—the cutting and burning of trees to convert land to grow crops, extract oil, and raise livestock—which he said accounted for 20 percent of Earth's human-caused carbon emissions. Schwarzenegger portrayed cap-and-trade as the definitive weapon for helping developing countries such as Brazil and Mexico conserve their forests and reduce global greenhouse gas emissions. As part of his climate action plan, the governor endorsed the use of carbon offsets within a cap-and-trade system. Offsets would allow California's polluting industries to pay someone else anywhere in the world to reduce their emissions by engaging in activities such as forest conservation. Under this approach, carbon offsets allow pollution at home only if developing countries keep their forests intact and do not use their own natural resources. Industries in the United States are permitted to continue to pollute, most often in low-income communities of color, such as Richmond and Oakland.[2]

The potential use of Reducing Emissions from Deforestation and Degradation (REDD) carbon offsets, moreover, has been a contentious issue in international climate talks. Some Indigenous rights leaders have argued that offset programs shift the burden of mitigating pollution to the Global South, where the commodification of tropical forests as carbon sinks can result in the displacement of Indigenous communities. Despite these purported conflicts, the governor soldiered on and pushed for an international agreement that committed California, along with 35 states and provinces (located in Brazil, Colombia, Indonesia, Ivory Coast, Mexico, Nigeria, Peru, Spain, and the United States) to establishing the Governors' Climate and Forest Task Force. The task force's mission is focused on advancing REDD programs throughout the globe. At the signing ceremony, before scores of reporters and news cameras, Schwarzenegger proclaimed that with the agreement, the signatories were targeting their collective forces on the climate problem and requiring their states to jointly enact rules, incentives, and tools to ensure reduced emissions from deforestation and land degradation.[3]

When I looked around the lavish international ballroom, at the crystal chandeliers and multimillion-dollar lighting and audio systems, I noticed something was missing. Indigenous leaders from the Global South were nowhere to be seen. They were not a party to the agreement, and their voices were not represented in the global climate change negotiations. Their exclusion occurred despite Schwarzenegger's insistence that fairness and equity were key to these discussions and that he wanted to see that no one was held back in combating the global climate change crisis. On one level, the contradiction was perplexing. At the same time, the climate summit reminded me of the imperialist fantasy represented by *Predator*. Against the backdrop of untamed nature, a white male action hero fights to save humanity, and Indigenous inhabitants—if their existence is even acknowledged—are seen as subversive. At the Beverly Hills Hilton, a privileged space of wealth and power, there seemed to be echoes of this trope in the forest offset deliberations. Global elites were shaping forest protection as they liked and were imposing a single, dominant narrative for saving the world.[4]

Less than two years later, the climate action governor convened another global meeting of policymakers to negotiate a new plan, a memorandum of understanding with the states of Chiapas, Mexico, and Acre, Brazil. The agreement outlined goals for linking the world's first state-to-state forest offset program and created the REDD Offset Working Group to develop recommendations for implementation. Conspicuously absent again were the voices of Indigenous communities from the Global South, whose culture and livelihoods depend on the forests now viewed as tradeable carbon market commodities. Nevertheless, this agreement would set a powerful precedent for connecting offset programs in the Global South with carbon markets in industrialized states. Chiapas and Acre planned to supplement these efforts with the adoption of various REDD-type land readiness programs, anticipating a connection with California's cap-and-trade program.[5]

Such preparation efforts have provoked discontent from both Indigenous rights leaders in the Global South and California environmental justice advocates. In order for environmental justice priorities to be incorporated into state-level climate change policy, shared ideas of equity and fairness first have to be negotiated among parties with differing initial preconceptions. This was difficult enough when all parties involved were within the state of California. But international offsets present even greater challenges. Because advocates' notions of justice are rooted in local, embodied experience, a system that can potentially cause inequities in distant and very different locations meant environmental justice groups had to adopt new strategies. In response, these groups forged a translocal justice movement and argued that such offsets can produce pollution hot spots within California's disadvantaged communities while facilitating possible human rights violations in Indigenous forest communities. The conflict over offsets has seen international delegations of Indigenous leaders from Brazil, Mexico, Ecuador, and Nigeria joining California environmental justice advocates in Sacramento to oppose the program.

These translocal coalitions have built links between local struggles, international organizations, and global climate change regimes. Anthropologists describe translocal social movements as engaging in "socio-spatial dynamics and processes of simultaneity and identity formation that transcend boundaries." In this sense, these translocal coalitions can be examined as configurations of place-based advocacy campaigns that exchange ideas, knowledge, practices, materials, and resources across disparate sites to enact social change in relation to climate change.[6] Human geographers further describe these social movements as occupying "enduring, open, and non-linear processes, which produce close interrelations between different places and people." These interrelations and forms of exchange are created through open flows and networks that are "constantly questioned and reworked."[7]

Networked communication and collaborative actions have amplified these groups' voices, thus enabling them to influence climate change policy within and beyond the nation-state. In this chapter, I describe how and why this social environmental movement developed at the intersection of overlapping scales (cities, regions, and nation-states), each of which provides opportunities and limitations for climate action and justice. Disagreements over the best course of action to address climate change do not concern just a single place or scale. Rather, each proposed solution is subject to multiple perspectives, held by people with disparate histories, levels of access to political power, and knowledge production practices. I argue that for California's climate change policy to be successful, it must incorporate marginalized voices and diverse perspectives from within the state and around the globe. This chapter calls out the high stakes of expanding California's carbon market to the Global South and raises the important question of who has the power to protect nature and humanity from the existential threat of climate change. Moreover, I highlight how the conceptualization of environmental justice as an organizing theme has spread horizontally throughout California, as well as the methods through which it travels, is adapted, and is now vertically linked to the Global South. Environmental justice advocates and Indigenous groups are thus building power in relation to global climate change policy in an informed and effective way. Crossing boundaries of scale and distance, the campaign against REDD offsets can serve as a model for future climate action.

Emergence of a New Translocal Advocacy Network

While policymakers in California proclaimed the benefits of forest offsets at international events, several Indigenous groups from the Global South protested the lack of consultation during the development of a proposal that could impact their lands and livelihoods. On September 26, 2012, over 40 Indigenous protesters from the Lacandon Jungle and members of international nongovernmental organizations gathered outside the Governors' Climate and Forest Task

Force meeting in Chiapas, Mexico. The meeting brought together government and business officials dedicated to implementing REDD policies globally. They represented 16 local governments of six countries (Mexico, Brazil, Indonesia, the United States, Peru, and Nigeria) that between them held 15–20 percent of the world's forests. Schwarzenegger was not in attendance; he no longer served as California's governor.[8]

Among the most vocal protestors at the task force meeting was Eufemia Landa Sanchez, a Tzeltal Mayan from a remote corner of Chiapas, adjacent to the Montes Azules Biosphere Reserve, one of Mexico's largest protected forests. Denied a chance to address the meeting, Sanchez seized the microphone during the open plenary and spoke before a packed auditorium of several hundred participants: "We have come before you today to denounce the programs and projects that threaten to dispossess us of our territories and our resources; programs that bad governments have attempted to impose for a long time; now they have a new pretext: climate change and the project they call REDD. . . . Why don't they consult us? Why do the wealthy want to impose their will by force? The jungles are sacred, and they exist to serve the people, as God gave them to us. We don't come to your countries and tell you what to do with your lands and your lives. We ask for the same respect."[9] Sanchez's denunciation of offsets invoked the nature of the forest as a home, as historically contested territory, and as sacred space. Her intervention highlights how policy and action on climate change map onto a range of existing meanings and dynamics concerning political power and the ownership and use of land.

For these reasons, according to Jeff Conant, an advocate with the international nongovernmental organization Friends of the Earth and author of *A Poetics of Resistance: The Revolutionary Public Relations of the Zapatista Insurgency*, Chiapas is the wrong place to test a new market-based mechanism from California. The southernmost of Mexico's 31 states and the birthplace of the 1994 Zapatista rebellion, Chiapas borders Guatemala and is the nation's poorest and most Indigenously populated state. Chiapas has a long history of conflicts over land tenure, in particular in the Lacandon Jungle area, where Indigenous peoples have for centuries faced forced removal from their native territories by government and business interests.

In recent years, the government's forest conservation programs have begun to delimit "natural protected areas" in order to generate offset credits that can be linked to international carbon markets, such as California's cap-and-trade system. According to Conant, "What this means in practice is a mandate for . . . [landholders] to cease planting their traditional crops, which are seen as harmful to the jungle, and to increase patrolling of their territory against outsiders, designated as 'invaders.' Those invaders, generally speaking, are indigenous communities who have never had formal title to the land, but who have been settled in the region for hundreds, if not thousands of years."[10]

Groups such as Conant's Friends of the Earth and the Indigenous Environmental Network (an Indigenous human rights group) contend that offset programs are creating a "perverse economic imperative" that could lead to forced displacement of Indigenous communities in Chiapas and Acre. Some Indigenous activists see REDD as a direct threat—not only to native lands but also against forest-linked cultural practices and the ability of their communities to reproduce those traditions. According to David Schlosberg and David Carruthers, Indigenous leaders articulate "environmental injustice as a set of conditions that remove or restrict the ability of individuals and communities to function fully—conditions that undermine their health, destroy economic and cultural livelihoods, or present general environmental threats." Their opposition to offsets can thus be seen as an expression of political freedoms and agency concerning how the environment is protected and how traditional cultures and practices are preserved.[11]

Studies of forest carbon projects in the Global South have also demonstrated how forest carbon markets can alter property rights and allow financiers, multinational corporations, and conservation organizations to gain control over land management decisions while undermining the authority of local Indigenous groups.[12] Research by human geographers highlights the equity concerns for Indigenous peoples whose customary land tenure rights may be insecure and whose means of livelihood may be targeted as the source of forest degradation. For example, after the 2010 agreement was signed between California, Chiapas, and Acre, the government of Chiapas faced significant scrutiny. Critics claimed that the state's forest carbon projects excluded women and individuals without formal property rights and that the resettlement of some Indigenous communities was achieved through the withdrawal of state-supported medical services. Opponents of California REDD offsets see the program as fueling the demand for new offset projects that could intensify land disputes between Indigenous and non-Indigenous populations, possibly resulting in forced evictions and disruptions of small-scale forest uses such as wood gathering, hunting, and fishing. Environmental groups such as Greenpeace also claim that international offsets could threaten biodiversity and native ecosystems through the conversion of natural forests into "carbon plantations," or less biologically diverse ecosystems, through afforestation and reforestation strategies.[13]

At the same time, steps toward an international carbon offset market have also been contested by California environmental justice advocates, whose focus is on the local effects of air pollution. These activists, who take a "California first" perspective of AB 32, oppose linkage programs because they may weaken requirements for direct emissions reductions within the state (fig. 6.1). They have pointed out, for example, that the Shell Oil Company (the US subsidiary of Royal Dutch Shell), which owns two oil refineries and numerous other high-polluting facilities in California, was allowed to purchase 500,000 forest offsets to meet its AB 32 obligations. Rather than reducing emissions near any of the company's California facilities, these offset credits support efforts to preserve a 200,000 acre forest 2,400 miles away in Michigan's pristine Upper Peninsula. The offsets are estimated to produce millions in potential gross compliance cost savings for Shell. Polluting industries support forest offsets in the Global South because they represent a more affordable option for reducing carbon emissions when compared to domestic mechanisms. Activists, however, argue that to effectively combat global climate change, policymaking must consider the phenomenon's "interconnectivity of economic, political, social and ecological" systems across temporal and spatial dimensions (also referred to as systems thinking or a feedback loop approach). For example, they claim that US imports of crude oil from the Amazon are driving the destruction of some of the rainforest ecosystem's most pristine areas and releasing large amounts of greenhouse gases. Activists point to an Amazon Watch study that shows that American refineries process

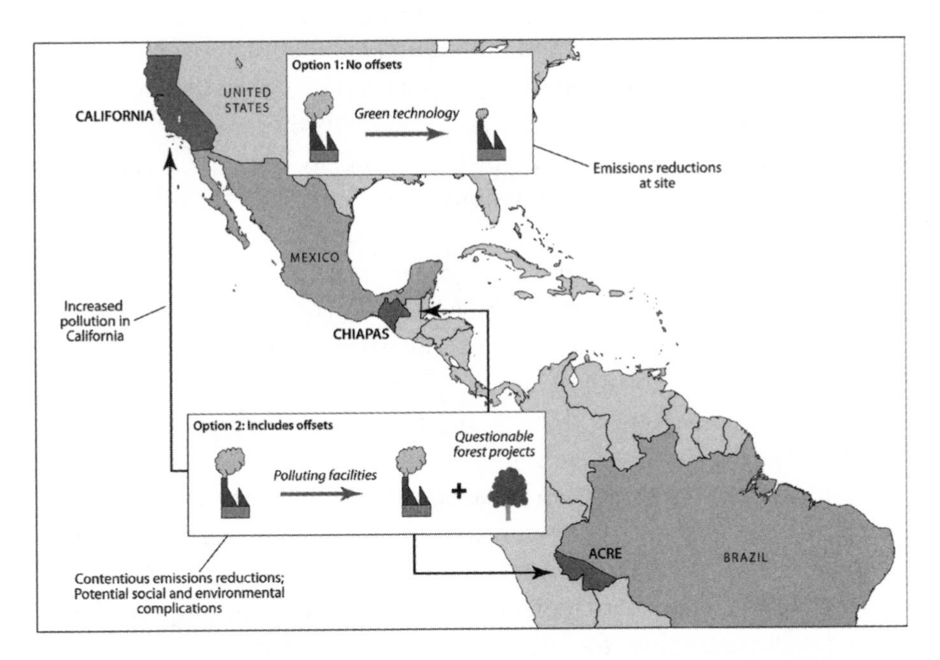

Figure 6.1. Overview of environmental justice groups' opposition to international forest offset projects. Adapted from Greenpeace/Sue Cowell

over 230,000 barrels of Amazon crude oil a day. California's refineries are responsible for a significant majority of this amount; they produce an average of 171,000 barrels (7.2 million gallons) a day, comprising 74 percent of all Amazon crude oil imports to the US.[14]

As California attempts to expand its climate change program beyond its borders, the issue of forest offsets encapsulates the conflicts between global and local perspectives on climate action. A central question is the extent to which polluters are responsible for reducing greenhouse gas and co-pollutant emissions in their local communities. According to Guillermo Mayer, president of Public Advocates (a nonprofit law firm and social justice advocacy organization), offsets decrease the likelihood that California residents who are disproportionately burdened by air pollution would benefit from the state's climate change laws: "Instead of reducing the pollution locally through better technology upgrades or ramping down emissions, [polluters] get to buy trees in another part of the world. The residents nearby aren't helped."[15] Environmental justice advocates argue that in terms of social equity, rather than market finance, offsets are an inefficient method for polluters to meet a portion of their emissions reduction targets because environmental and health benefits are being exported out of state. They further claim that California has no international regulatory authority or political capacity to enforce forests offset regulations to prevent harm to distant populations and landscapes.[16]

Forests thus have heightened significance in global climate change debates. Not only do they fit the conventional criteria for environmental protection, but also, by simply existing, they affect calculations about net global greenhouse gas emissions and can therefore be seen as enhancing value in carbon markets. Political ecologists have argued that forest carbon offsets have abstracted nature for the realization of "fictitious capitalist value." The production of technical reports and measurements and the issuance of certified offset credits allow the next 100 years of forest carbon storage to be realized today. The offset credit serves as a link between the trees' use and exchange value, for it is the credit and not the actual carbon stored in the ground that enables industries to meet regulatory obligations while continuing their polluting activities. Other scholars argue that forest offsets exist to transform a future goal of climate action into an already accomplished present. These accounting methods and the outsourcing of emissions reductions are the basis for environmental justice groups' opposition to offsets. They have long held that offsets serve as illusory promissory notes of carbon mitigation that allow polluters to eschew responsibilities to the local communities that are harmed by their activities.[17]

These contentious debates have translated geographically distant groups' concerns over the use of offsets into a new translocal environmental justice movement. Along with networking through introductions by advocates at Friends of the Earth and the Indigenous Environmental Network, digital technology has enabled California environmental justice groups to communicate with Indigenous leaders in the Global South and to link their individual concerns into a multiscalar narrative of opposition to REDD. Many of these groups' equity claims focused on the ethical importance of the United States taking historical responsibility for greenhouse gas emissions reductions and the potential for offsets to result in land dispossession and human rights violations. In this sense, we see how diverse notions of environmental justice travel, evolve, and are horizontally and vertically linked.[18]

For California advocates, there has been a growing need to develop a global consciousness in the environmental justice movement. While their focus has always been local, they recognize the importance of connecting a local agenda with trends they see nationally and internationally around climate change. According to Mari Rose Taruc of the Asian Pacific Environmental Network, this realization began when the group's Native American allies, who represent sovereign nations, were participating in United Nations debates on climate change policy: "They were connected with other Indigenous groups from around the world and linking some of the threats of climate change but also exposing these false solution programs like forestry offsets and the Clean Development Mechanism Fund." Taruc explained that Native American advocacy groups, such as the Indigenous Environmental Network, developed campaigns that brought in US environmental justice groups to weigh in at the UN and other global spaces. This collaboration was solidified in 2009 and 2010 at meetings where advocates (including Taruc) gathered for the UN climate talks in Copenhagen, Denmark, and Cancún, Mexico.[19]

In these arenas, environmental justice groups learned how linking their local air quality campaigns to help clean up greenhouse gas pollution can also aid Indigenous communities in the Global South. For Taruc, it was a hostile incident that would solidify her opposition to offsets: "I was holding up a NO REDD sign in Cancún, Mexico, at the United Nations Conference of the Parties. And for holding up that sign, I was detained and tossed out of the climate negotiations by the United Nations security. And you would think I would be afraid after that experience. But actually, I was encouraged. Because to the right of me at the protest was the president of Bolivia and Indigenous supporter, Evo Morales. And to the left of me were leaders of social movements from the Western Hemisphere, from the MST [Landless Workers' Movement] of Brazil and the Vía Campesina of Mexico. . . . And in front of me were Native American brothers and sisters from the Indigenous Environmental Network, who have been campaigning to end REDD offsets." This contentious encounter made it more important for Taruc to take what she was learning in these global spaces back to California. And in her own words, it was "with that spirit that we as an environmental justice movement have been opposing REDD offsets ever since."[20]

Interactions at global forums such as the United Nations between Indigenous human rights groups and California environmental justice groups laid a strong foundation to develop new translocal strategies against offsets. Working together in person and via conference calls and email during 2011 and 2012, these groups organized an opposition campaign. They developed a joint narrative that explains how mechanisms that pay countries and landowners to save forests could hurt those who live among the trees and next to polluting industries. These encounters led to a common understanding between diverse groups about the spatial implications and global reach of California's carbon market. More than 30 Latin American and US organizations drafted and distributed a collective anti-REDD statement to policymakers at the forest task force gathering. Among the signatories were Oakland-based environmental justice groups, including Communities for a Better Environment and Movement Generation.[21]

According to Donatella della Porta and Mario Diani, such network-building exercises "contribute both to creating the preconditions for mobilization and to providing the proper setting for the elaboration of specific worldviews and lifestyles."[22] The anti-REDD social movement, moreover, illustrates the ways in which California activists developed connections with groups outside their everyday networks. The sociologist Mark Granovetter argues that interaction among organizations with few common interests, or "weak ties," enables increased possibilities for the circulation of new ideas and strategies. He asserts that such interactions are often more important than those among groups with "strong ties," because frequent engagement may inhibit innovations in strategy and the inclusion of new ideas.[23]

This perspective supports Margaret Keck and Kathryn Sikkink's seminal analysis of translocal advocacy networks, which they depict as "vehicles for communicative and political exchange, with the potential for mutual transformation of participants."[24] In the predigital era, connecting the efforts of geographically distant grassroots movements —even those with similar interests—was often difficult. In our contemporary world, however, the logic of networks provides a kind of connection that builds collective power without erasing the specificity of each networked location. The use of digital communications enables rapid linkages so that activists and nongovernmental organizations can pursue a visual and symbolic politics where the human scale of carbon markets is made evident.[25]

The Indigenous rights protest at the forest task force meeting revealed a dense web of south-north exchange (fig. 6.2). The existence of these networks meant that governments promoting forest conservation projects that involved perceived harm to Indigenous communities could no longer control information flows. Although task force participants acknowledged the opposition to offsets, they seemed to underestimate the depth of the conflict. William Boyd, the task force's senior advisor, stated that "any broad public policy is going to generate opposition. We understand that, and we see the need to do a better job at communicating our objectives."[26] In response, protesters developed alternative narratives in forums and workshops during the weeklong meeting, as well as protest marches through the city center of San Cristóbal de las Casas, Chiapas. These activities, which were recorded and uploaded to social media platforms, amplified the message of opposition to offset programs. The protesters depicted opposition to the program not as a result of

Figure 6.2. Horizontal Activism: Indigenous protest at the 2012 Governors' Climate and Forests Task Force meeting. Courtesy of Jeff Conant

a lack of scientific understanding but rather as founded on networked and embodied knowledge about the potential negative effects of global climate change policy in the tropical forest and the urban street.[27]

Translocal Justice Movements at the Capitol

Three weeks after the forest task force meeting, an international delegation of Indigenous leaders from Brazil, Mexico, and Ecuador traveled to Sacramento to register their opposition in person at a California Air Resources Board public hearing on October 18, 2012. The delegation was hosted and sponsored by Friends of the Earth and several California environmental justice organizations, such as, the Asian Pacific Environmental Network, the California Environmental Justice Alliance, and the Center on Race, Poverty, and the Environment. Its international members included Ninawa (José Carmelio Nunes) of the Federation of the Huni Kuin people in Acre, Brazil (a western state bordering Bolivia and Peru); Rosario Aguilar, a public health worker with T'sunel Bej (a Lacandon Jungle–based health collective); Berenice Sanchez Lozada of the Global Alliance of Indigenous Peoples and Local Communities against REDD and for Life; Gloria Ushigua of the Association of Zápara Women; Marlon Santi of Sarayaku and Ecuador's Indigenous movement; and representatives from the Indigenous Environmental Network.[28]

While digital communications aided the development of a south-north exchange, advocates understood that physical sites of assembly were still the most effective way to collectively express resistance, demonstrate the strength of a movement, and challenge the dominant power. California represented the locus of power, as it is the only local jurisdiction in the world that is considering REDD offsets as part of a cap-and-trade system.[29] The visibility of the international Indigenous delegation enabled California policymakers to witness the purported human-scale impacts of carbon markets and see beyond the notion of tropical forests as mechanisms to generate economic value and environmental services. The presence of Indigenous leaders alongside California environmental justice advocates, moreover, was a poignant display of symbolic politics. It called upon a narrative of the potential south-north environmental injustice derived from California's global, market-based climate solutions.

The international delegation to Sacramento, however, did not come together easily. In preparation, participants held several in-person organizing meetings in California to exchange understandings of the state's complex climate policy landscape. Initially, Indigenous rights groups were apprehensive about joining environmental justice groups in Sacramento, in part because they felt that those groups might stand to benefit from offset revenues at the expense of Indigenous communities. One environmental justice advocate recalled, "I remember being in a meeting where they were pissed at us for even thinking about benefit sharing—such as using revenue from the Climate Change Community Benefit Fund [SB 535] to invest in environmental justice projects in our communities" (see Chapter 5). In these meetings, California environmental justice advocates had to earn the trust of Indigenous delegates by explaining how their principles were aligned with Indigenous rights and that activists were fighting on multiple fronts while engaging in a complicated mix of climate change programs: "To explain to them California's complicated system was hard. Because they saw that we were taking money from the climate investments via SB 535 and they knew the threat of that. There was a feeling that we were going to sell them out on offsets and carbon trading."[30]

With close ties to both the Global South and California advocates, the Indigenous Environmental Network facilitated these initial discussions and helped to open communication and build trust. Through this process, environmental justice groups detailed their efforts to oppose cap-and-trade, including their lawsuit against the state of California to stop implementation, and clarified that SB 535 funds did not include revenue from offsets. This sharing of history and goals enabled participants to reach some common ground and develop an opposition campaign. Their core argument was that California represented one of the largest markets for offsets and that collaboration was crucial to ensure REDD did not proliferate globally.[31]

As a result of their organizing, the translocal coalition of Indigenous rights groups and California environmental justice advocates displayed a united front at the board hearing in Sacramento. They expressed support for California's efforts to reduce greenhouse gas emissions and agreed that conserving tropical forests was critical to protecting the global climate. However, they also warned that efforts such as REDD ignored or harmed a range of needs, including land tenure, small-scale economic activities, and a variety of cultural practices. While an international forest offset program had not been finalized, they argued, the agreement that Governor Schwarzenegger had signed set a precedent that was already leading to the eviction of Indigenous peoples from their lands in Chiapas and Acre in preparation for a California-linked REDD program. The coalition further claimed that REDD readiness programs were undercutting Indigenous community efforts to gain formal land tenure.[32]

Activists presented stories of purported harm already caused to community members by these REDD-related measures. In her testimony to the board, Rosario Aguilar described the steps that the state of Chiapas was taking to implement REDD readiness programs. These included the alleged relocation of "illegal groups" that were deemed to be unlawfully occupying natural areas in El Triunfo Biosphere Reserve and Cañón del Sumidero National Park. Presenting a brochure distributed by the state of Chiapas during UN climate negotiations, Aguilar declared, "This official document . . . clearly proves that REDD results in evictions. In fact, they brag about having evicted 172 communities to do REDD."[33] Describing the experiences of Doña Juanita, one of the elders who founded the community of Amador Hernández in the Lacandon Jungle, Aguilar also claimed that, due to a lack of government-supported medical services, Doña Juanita developed foot gangrene that spread and eventually required amputation of one of her legs. Aguilar stated that the Chiapas State government was suspending essential medical care to remote Indigenous communities in an attempt to pressure villagers into relocating to larger urbanized areas to allow offset projects to proceed.[34]

Following Aguilar's statement, Ninawa addressed board members. Ninawa testified that the formal agreements between Acre, Chiapas, and California were signed with no direct consultation with the Indigenous people in those regions. He referred to the UN Declaration of the Rights of Indigenous Peoples, which guaranteed the right to free, prior, and informed consent in the development of projects such as forest offsets. He testified that in Acre, the demarcation of Indigenous territories was "paralyzed because they want to take our lands to make profit from environmental services through programs like REDD. We will not and cannot trade our hunting, our fishing, and our lives for pollution. You cannot trade pollution for nature." Ninawa further urged the board to reject international offsets in the climate change program because these projects were already affecting Indigenous livelihoods and access to ancestral lands: "These are lands that are being included in REDD projects and [that] large companies, operators, and landholders have had an eye on. They are restricting our way of life and our ability to have access to our traditional hunting, fishing, and gathering sites. So for this reason, we are urging you not to accept REDD in your trading program."[35]

Following the testimony of more Indigenous leaders, California environmental justice advocates proclaimed their solidarity with Indigenous peoples in the Global South. They reinforced the narrative of offset programs facilitating land dispossession and government repression. Advocates pressed board members to abandon international forest offsets to prevent an onslaught of human rights violations abroad. They argued that the integrity of California's climate change programs could otherwise be threatened and urged policymakers instead to focus their energy on reducing direct emissions within California. In the words of Sofia Parino, senior attorney with the Center on Race, Poverty, and the Environment, "We stand with our international brothers and sisters. We believe [that] REDD programs . . . are bad for communities internationally that are being decimated from the program and [California] communities that are not receiving the benefit from local pollution reduction."[36] When the board hearing concluded, members of the translocal coalition continued their lobbying efforts at the state capitol (fig. 6.3). They met with senior staff from the offices of the governor and state legislative leaders, whom they presented with a letter signed by more than 30 California-based organizations opposing REDD. During these lobbying meetings, the coalition informed policymakers of what they believed to be the inherent problems with offsets, as well as problems specific to the California-Chiapas-Acre context.[37]

Figure 6.3. Vertical Activism: Coalition of Indigenous and environmental justice groups lobbying at the California State Capitol in Sacramento. Courtesy of Jeff Conant

Following the capitol lobby day, the coalition took strategic advantage of several converging opportunities to organize a No REDD Tour throughout Northern California. Coalition members sought to educate the news media about the effects of offset programs on people living in both industrialized and forest regions. The coalition, which viewed California as being at the forefront of the global offsets debate, presented the perspectives of several leaders of forest peoples, who assessed the impacts of REDD. They used speaking events, film screenings, and meetings between Indigenous leaders and Californian allies to raise public awareness. The No REDD Tour coincided with meetings held by the REDD Offset Working Group and the annual Bioneers Conference. Advocates envisioned the tour as an opportunity to assemble an important constituency in the offsets debate that could facilitate a global platform for Indigenous and environmental justice leaders.[38]

The tour led to a larger policy discussion in the capitol and news media in subsequent months over California's ability to monitor the integrity of international offsets in the face of perceived ongoing corruption in developing countries. For example, the legal scholar Alan Ramo noted that any international offset program implemented in a developing country would depend upon the host country or third parties for validation.[39] Corruption at any stage, including initial reporting, verification, and monitoring, could undermine offset programs. Ramo's comments, influenced by the visibility of the translocal coalition, further raised concerns with capitol staffers. These included whether the board could monitor international offsets in the same manner as domestic ones and whether California should entrust countries facing high levels of perceived public sector corruption with the responsibility of validating offsets. The board, unlike the federal government, lacks international authority to enforce AB 32's provisions or intervene in another country's sovereignty. Several senior capitol staff members said that California should be cautious about developing linkages where doing so could induce or exacerbate human rights violations in the Global South.[40]

Research by the forest scientist William Sunderlin and others has also questioned the viability of subnational forest offset initiatives. They found, in a sample study of 23 REDD projects in six countries (Brazil, Peru, Cameroon, Tanzania, Indonesia, and Vietnam), that six projects had closed, four no longer labeled themselves as REDD, and only four were selling carbon credits on the voluntary market. Some projects in Tanzania, Indonesia, and Vietnam closed because they had been unable to sell forest carbon credits before seed funding from the government of Norway ran out; others expired due to local changes in political leadership that undermined support for REDD programs. Sunderlin's group concluded that a lack of stable financing, technical capacity, and secure land tenure all meant that subnational REDD projects would require a binding international agreement to be successfully implemented.[41] A conservation consultant with the US Agency for International Development (USAID) further argued that local communities on the frontlines of policy and project implementation could disproportionately bear the potential costs of unsuccessful REDD projects. Should international offsets fail, he contended, Indigenous communities' livelihoods could suffer from the negative impacts of changed resource and land use patterns. That said, private sector investors supporting a California offset linkage program might also suffer economic losses.[42]

Some of these concerns seemed validated when, in 2012, the board voided nearly 90,000 offset credits generated by Clean Harbors Inc., the nation's largest incinerator of chlorofluorocarbons (typically found in refrigerants such as Freon). While the burning of chlorofluorocarbons is banned in California, it is allowed in Arkansas as a carbon offset. Clean Harbors' incinerator is sited in El Dorado, Arkansas, a town with some of the state's worst air quality and where one-fourth of residents live below the poverty line. California's cap-and-trade program authorized the board to invalidate previously issued offsets if the relevant facility had failed to comply with environmental, health, and safety requirements. The board voided Clean Harbors' offset credits after it was penalized by the US Environmental Protection Agency for improperly storing and disposing of hazardous waste and failing to comply with air emissions standards.[43]

In invalidating Clean Harbors' credits, the board did not identify offset purchasers. Disclosing such information is prohibited by law; each company's strategy in purchasing offsets is considered proprietary, market-sensitive information. However, the law firm Latham & Watkins, which represents some polluters, estimated that nearly 20 California companies invested in Clean Harbors. All were required to purchase additional credits to compensate for the cancelled offsets.[44] According to California environmental justice advocates, the Clean Harbors case represents the dangers of issuing offset credits for distant facilities instead of requiring direct pollution reductions within California. The violations may have produced poor air quality for low-income Arkansas communities while increasing businesses' compliance costs. If the board cannot effectively monitor US domestic offset operators that violate environmental safety and health laws, activists argue, how will it validate offset programs and avoid human rights violations in the Global South?[45]

The Boomerang Effect of Translocal Coalitions

Throughout 2012, California environmental justice advocates and Indigenous leaders demonstrated how powerful networked coalitions could shape the social and environmental narratives of carbon reductions. During the next four years, the translocal coalition helped stall momentum toward an international forest offset linkage program. Most importantly, its protest created additional market uncertainty. According to the carbon market trader Harold Buchanan, a number of technical and legal issues and political hurdles arose that together appeared insurmountable: "The complexity of getting international credits of these types into a state-based system is overwhelming, not to mention the lack of popularity. There are certain constituencies that do like it, but frankly there are many, many more that see the problems and will pull out all the stops to stop the process. I don't think it's realistic."[46] Other carbon market consultants also expressed doubts that California would take on international offsets.[47] Although California's cap-and-trade regulations include a placeholder to allow international offsets to enter its program, the California Air Resources Board is taking a cautious approach to adopting them. In fact, after the No REDD Tour was launched, the board seemed to distance itself for a while from international offsets. According to a board spokesman, "We have [a memorandum of understanding] to observe development of sector-based [forest] projects in Chiapas and Acre, but no agreement to accept those projects. At this point we have just issued the first offset credits for domestic projects, and that is our primary focus right now."[48]

The potential for the translocal coalition to influence the outcome of California's global offset program is consistent with Keck and Sikkink's seminal research, *Activists beyond Borders*. This work offered the field of human rights history a model of the new phenomenon of global human rights activism. What the authors termed the "boomerang pattern" can be observed in the way that nongovernmental organizations in the Global South work with similar domestic groups in the Global North to address human rights violations in their own countries (fig. 6.4). In the case of translocal opposition to REDD, Indigenous groups in Chiapas and Acre tried to petition their own governments (state A) to stop the development of offset programs. However, as described by David Bassano, the relative weakness of civil society in places such as Mexico and Brazil meant that the groups lacked sufficient power to influence their governments and could even become targets of repression themselves—a situation diagrammed as the "blockage" in the pattern. The groups therefore connected with California organizations for assistance.[49]

The Global North organizations, in comparison had greater free-

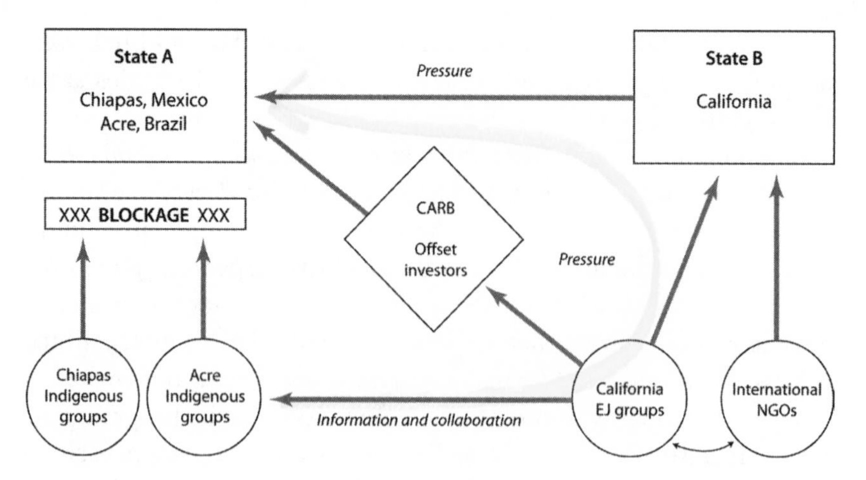

Figure 6.4. The boomerang pattern of REDD. Adapted from Keck and Sikkink, *Activists beyond Borders*, 13

dom of action and benefited from stronger civil societies. Information sharing and collaboration between organizations in both regions led California environmental justice groups to petition their government (state B) to block the linkage program. Such pressure also affected decision-making in Acre and Chiapas (state A) because the programs could only succeed if California financed and adopted the global offset market and governance structures. Through direct lobbying, media, and protest, the translocal coalition attempted to dissuade California policymakers from linking its cap-and-trade program to the Global South, with the goal of alleviating some of the purported oppressive effects of REDD-related programs. California policymakers, perhaps not wanting to be viewed as enabling policies that could provoke human rights abuses abroad, began to slow down the process of approval of global offsets.[50] Members of this powerful coalition were able to "jump" from the local scale (the tropical forests of Chiapas and Acre) to the global scale (meetings of the board and the forest task force), thus affecting the scope of climate change policies.

According to the sociologist Brian Gareau, tracing the scalar debates and political action over issues such as offsets also reveals the ways in which environmental governance often ignores society-ecology relations and attempts to treat certain aspects of "the environment" in isolation. For instance, greenhouse gas mitigation is treated as separate from the public health of local communities, and tropical forest conservation as separate from Indigenous livelihoods: "We are reminded here of how local, national, and international scales are established in relation to one another. The socio-ecological conditions existing in and around California . . . extend from the local to the global scale, and the process by which powerful actors shape those arrangements changes as they move across scales."[51] Through the actions and narratives of the translocal anti-REDD coalition, it becomes clear that policies to address climate change are embedded within networked configurations that extend from the local context to global relations.[52] This multiscalar framework for evaluating the fairness and effectiveness of climate action has begun to shape the broader system of decision-making in Sacramento.

From Tropical Forests to Short-Lived Climate Pollutants

By creating a perspective on the local effects of policies formulated on the global scale, the translocal coalition has begun to bring the legitimacy of California's carbon policies into question while suggesting ways to ensure their accountability. The boomerang pattern is a useful tool for understanding these dynamics; however, its return journey is often not straightforward. Because translocal environmental justice movements take place at different scales, proposed solutions are subject to negotiation by a collection of individuals and organizations with different levels of power, knowledge production practices, and histories. As a result, the boomerang can create discrete political outcomes that benefit some coalition members more than others. In the case of the anti-REDD coalition, only part of its proposals would eventually become law, because of significant disagreements among traditional environmentalists about the importance of offsets and the decision by California's largest Native American tribe to develop forest offsets on their sovereign lands.

These political dynamics came into play on February 22, 2013, when environmental justice groups joined Senator Ricardo Lara, an East Los Angeles Democrat, to introduce SB 605, a measure that would limit the development of offset projects outside California. In supporting the bill, environmental justice groups sought to prioritize greenhouse gas reduction projects that reduced co-pollutants within California while preventing alleged human rights violations in the Global South. Breaking ranks with the environmental justice movement, however, the Nature Conservancy and the Environmental Defense Fund opposed the legislation. In the last 10 years these organizations, embodying a worldview in which international offsets generate win-win outcomes, have consulted on or sold voluntary offset projects to polluters across the globe. They view REDD as a means to tackle climate change and promote sustainable development in local Indigenous communities and as a financial vehicle to extend their organizations' global reach in forest conservation. To them, California is a high-stakes battleground for REDD. Success, in their perspective, could provide valuable proof of concept to inspire other governments around the world to adopt similar programs.[53]

These traditional environmental organizations, moreover, often prioritize the global nature of greenhouse gases and dispute the argument that their reduction should be linked to the mitigation of local co-pollutants. According to political ecologist Kathleen McAfee, their support of offsets is based on the principle that all emissions reductions, irrespective of where they occur, are tradable or equivalent in ecological terms. These groups argue that offset credits are beneficial because they provide polluters choice in a competitive national (and potentially global) market for the most cost-effective methods to reduce net greenhouse gas emissions while investing in conservation projects. By focusing on such global approaches to climate change, McAfee contends, they are promoting the "selling of nature to save it."[54]

This type of relationship between the private sector and biodiversity conservation organizations is becoming more prevalent with the emergence of narratives of the green economy. A 2011 study of the Nature Conservancy found that 15 of the organization's 26 board members are or have been directors of large transnational corporations; meanwhile, the same study found that more Nature Conservancy program directors had a corporate management background than training in biology or related fields. The study concluded that increased corporate involvement has coincided with increased promotion of market-based solutions to conservation problems.[55]

The increasing market-based orientation of traditional environmental organizations has compounded the tension between the local and global models for evaluating the effectiveness and fairness of environmental policies. This tension has been central to public debate over the future of California's climate change policy.

At the Assembly Natural Resources Committee hearing for SB 605 on August 13, 2013, the ornate setting and the neoliberal perspectives on display recalled Governor Schwarzenegger's convening at the Beverly Hills Hilton five years prior. The rallying calls for a global mentality for climate action conveyed the power relations represented in the committee room. During the hearing, the Environmental Defense Fund's lobbyist testified against SB 605. He argued that the measure would impede efforts to expand international offset programs and the ability to gain approval of voluntary offset projects under California's cap-and-trade program, all while excluding California from a global dialogue on climate change. The lobbyist stated, "The current version of the bill cuts off important opportunities for investments in projects that can stimulate reductions, not only in California . . . but in other nations as well. Cutting off opportunities like international forestry leaves California outside of the realm where we can help to participate and influence in a global dialogue. . . . So we just cannot support a bill that keeps California outside of that conversation."[56]

While the committee room represented a locus of power, one difference stood out. Unlike the 2008 summit at the Beverly Hills Hilton, in the SB 605 hearing, advocates speaking for disadvantaged communities in California and abroad had a prominent stage. More than two dozen environmental justice, health, and conservation advocates lined up to urge support for the bill. In particular, environmental justice groups' testimony discounted the Environmental Defense Fund's perspective. They countered that their communities would not benefit from such programs and might be harmed by them. Activists further claimed that neighborhoods near major polluters had historically not benefited from the state's green economy and that environmental protection investments had bypassed many of California's disadvantaged communities.[57]

The oil industry, the California Chamber of Commerce, the California Manufacturers and Technology Association, and carbon trading firms all opposed SB 605 from a financial standpoint. They developed economic models estimating that California's carbon market would have few compliance mechanisms, while businesses would face increased financial burdens in complying with greenhouse gas emissions reduction mandates.[58] Senior air board representatives have also adopted a market-based logic for climate policy; they assert that apart from reducing emissions, AB 32 is also intended to create an economic imperative for investment in new green technologies. According to a senior board officer, "You have to understand what we're doing here in California, in the national context. It's not just about emissions reductions. What we're talking about is a permanent shift toward a less carbon-intensive economy. . . . While there's an environmental imperative, there's also an economic imperative."[59]

During the SB 605 deliberations, such an economic imperative initially led the board to support recommendations developed by the REDD Offset Working Group. This group concluded that offsets in Chiapas and Acre would be more affordable than US domestic offsets and could ensure the stability of California's carbon market.[60] However, environmental justice advocates questioned the legitimacy of these recommendations and cited potential conflicts of interest in funding and the organization placed in charge of developing the report. The offset working group was led by the Green Technology Leadership Group, a nonprofit advocacy organization that represents the interests of both the suppliers and buyers of international offsets. The executive director, Tony Brunello, served at the same time as a principal at California Strategies, the second-largest lobbying firm in Sacramento. The firm's clients have included oil companies, carbon market traders, the offset operator Clean Harbors Inc., and the Nature Conservancy.[61]

Inherent in the SB 605 disagreement were differing ideas about what constitutes environmental protection—whether it solely concerns the preservation of "natural" areas and ecosystems or whether offset projects should be extended to include the urban environments and populations most affected by pollution. The debate over the appropriate policy scale to address the phenomenon of climate change has been further complicated by California's conflicting mandates and goals. AB 32 requires the reduction of global greenhouse gas emissions while providing opportunities to improve local air quality in communities already adversely impacted by air pollution (see Chapter 3). Alan Ramo therefore argues that the inclusion of out-of-state offsets may make that goal difficult to achieve: "The upshot is that while [the board] promised health benefits here at home from cap-and-trade, those may potentially prove illusory in exchange for the appearance of reductions elsewhere in the globe. Whether these offsets really swamp California's cap-and-trade program or simply spread the regulatory gospel around the world will be settled on the ground based upon the implementation of these programs."[62]

Despite significant opposition from powerful interest groups, Senator Lara succeeded in moving SB 605 out of the Assembly Natural Resources Committee. But the bill's chances of passage in the legislature were diminishing. This became more apparent in subsequent months, when Lara's mentor and the author of AB 32, former assembly speaker Fabian Núñez, now a partner at the high-powered lobbying firm Mercury Public Affairs, took an offset operator as one of his clients. This decision was coupled with the firm's announcement of its intention to open a satellite office in Mexico City to help Californian businesses navigate the Mexican government's trade regulations.[63] Such institutional and business connections among legislators, lobbyists, industry, and conservationists have often affected decision-making around environmental policy. As a former legislative leader and the author of AB 32, Núñez held significant influence. His involvement with offset operators sent a clear message to stakeholders that the cap-and-trade program should include out-of-state offsets.[64]

Another complication at the time came from the revelation that California's largest Native American tribe, the Yurok, was in the process of developing forest offset projects on its reservation near Redwood National Park. The board would issue the tribe more than 800,000 offset credits in one of the first forestry projects approved under cap-and-trade. Yurok officials stated that offset projects aligned with their goals to increase the tribe's land base and restore the forest near the Klamath River, in turn aiding the river's water quality and salmon fishery. Under offset rules, the Yurok can harvest some timber from offset project areas; however, they must factor such logging into their carbon sequestration calculations. The tribe has used the proceeds from offset sales to finance land acquisitions, including its 2011 purchase of more than 22,000 acres from a lumber operator.[65]

In 2013, the Yurok owned only half the land that constitutes their reservation; timber companies owned the rest. Tribal leaders intend to develop additional offset projects to purchase more land in and around the reservation. The Yurok Tribe has sold millions of dollars' worth of forest offsets to some of the state's biggest polluters, including major oil companies. According to the tribe's attorney, "sales are going on all the time," with industries looking to purchase more offset credits.[66] The Yurok Tribe has adopted a nuanced neoliberal argument that offsets can bring environmental, social, and financial gains. In the announcement of a new offset project, Thomas P. O'Rourke Sr., chairman of the Yurok Tribal Council, stated, "This carbon offset project will foster the restoration of a significant swath of forest. Our partnership with New Forests [an offset developer] will provide the Tribe with the means to boost biodiversity, accelerate watershed restoration, and increase the abundance of important cultural resources like acorns, huckleberry and hundreds of medicinal plants that thrive in a fully functioning forest ecosystem."[67] New Forests would go on to describe their partnership as a form of "economic diplomacy for climate mitigation" which shows that society values the ecosystem services of cap-and-trade.[68]

Offset projects on Native American forest lands would extend beyond California to include the White Mountain Apache Tribe of Arizona, the Joint Tribal Council of Passamaquoddy Tribe in Maine, the Confederated Tribes of Warm Springs Reservation in Oregon, the Round Valley Indian Tribe of California, the Mescalero Apache Tribe of New Mexico, and the Confederated Tribes of the Colville Reservation in Washington State.[69] Throughout the legislative fight over SB 605, the tribes remained silent and took no formal position on the legislation. Traditional environmental organizations and offset developers, however, asserted the social and economic benefits of forest offsets for Native American communities and used it as another basis for opposing SB 605.

Environmental justice groups have refrained from publicly rejecting offsets on Native American lands. When I asked an outspoken advocate with a Los Angeles–based environmental justice group about her perspective, she hesitated. This was a delicate and nuanced issue for her. While she understood the argument that offsets could help tribes make money, she still felt it was problematic from an environmental justice perspective. She told me, "So I go to first principles, and this is not right. Because you force people to fight for something that is inherently indefensible. It's like oil workers. Of course they want to keep their jobs. And of course they sometimes see us as enemies and accuse us of taking their livelihoods. But no, we want jobs that can send their kids to college and doesn't cost you your health and the health of our planet. So the offsets really bother me. It is sort of like splitting the baby, Solomon's choice. Which side of the baby are you going to take?"[70] Other environmental justice advocates, however, hold different perspectives. They do not consider offsets on Native American lands to be directly comparable with those proposed in the Global South and cite key differences in the colonial histories of the United States and Latin America. In the United States, treaties established a form of recognized sovereignty, which brought territorial boundaries and political influence that Indigenous groups in Mexico or Brazil often lack. These advocates argue that forest offsets provide Native American tribes facing limited economic opportunities with a means to achieve territorial integrity and improve the health of tribal lands and resources.[71]

By the time SB 605 reached the Assembly Appropriations Committee in late August 2013, it was clear that supporters of domestic and international offset projects had gained significant traction to stop the legislation. The committee held SB 605, deeming it too costly for the state to implement. The move delayed the measure until the following year, when Senator Lara could either amend SB 605 to address fiscal issues or let the legislation die. While the legislation was on hold, in December 2013, the Environmental Defense Fund arranged for a delegation of state legislators to visit Mexico to educate them about REDD-type projects and to showcase opportunities for linkage with California. The organization even paid to send Senator Lara along with the delegation. In addition to sponsoring such trips overseas, these groups were also involved with the forest task force and the REDD Offset Working Group. They consistently engaged with board staff, the governor's office, and the legislature to represent their interests in the development of an international offsets market.[72]

In subsequent months, Senator Lara and environmental justice advocates acknowledged that SB 605 had a slim chance of passage as written. While deleting all provisions relating to offsets, they amended the bill to require the board to look beyond just the seven global greenhouse gases regulated under AB 32 (see Chapter 1). The new amendments required the board to complete a comprehensive short-lived climate pollutant reduction strategy by January 1, 2016, that focused in particular on black carbon, a global and local pollutant (also referred to as a "super pollutant"). In the United States, black carbon is a major byproduct of diesel engine exhaust. It is emitted into the atmosphere as fine particles that remain for days to weeks. Black carbon is both a main contributor to climate change and a public health threat in disadvantaged communities.[73]

At the global scale, black carbon has been identified as the second-most important emission contributing to climate change, after carbon dioxide. It affects climate change by absorbing sunlight and reducing the cooling effect of reflective surfaces such as glaciers. In addition, black carbon harms plants when it lands on leaves by impeding photosynthesis. According to the board, depending upon geography, weather conditions, and time of year, black carbon has the potential to trap as much as 6,200 times more heat in the atmosphere than carbon dioxide over 20 years and 1,700 times more heat over a 100-year period. At the local scale, the US Environmental Protection Agency notes a link between black carbon exposure and a range of illnesses.[74] Increased black carbon emissions raise significant public health concerns because their fine particles can penetrate and lodge deep inside lung tissue, trigger cardiovascular and respiratory diseases, and contributing to premature death. Children and older adults are most susceptible to the negative health effects of black carbon, such as higher incidences of asthma or heart attacks. The US Environmental Protection Agency estimates that the average public health benefit of black carbon reductions in the United States can range in monetary value from $290,000 to $1.2 million per ton of particulate matter by 2030.[75]

To address the black carbon problem, SB 605's new amendments included the establishment of an emissions inventory, the identification of research needs and data gaps, and an analysis of existing and potential new control measures to reduce short-lived climate pollutant emissions. The measure also called for increased transparency in the formulation of the emissions reduction strategy through public workshops and inclusion of an environmental justice provision prioritizing measures that improve local air quality in disadvantaged communities. This provision required the strategy to identify disadvantaged communities using the state's newly developed, science-based tool, CalEnviroScreen (see Chapter 5).[76]

By the conclusion of the 2014 legislative session, Senator Lara and California environmental justice advocates could declare partial victory. Their pragmatic approach toward tackling global and local pollutants via SB 605 gained wide support and the signature of the governor, Jerry Brown. Although the provisions to prohibit international offset projects were removed from the bill, the initial introduction of an anti-REDD bill helped amplify public debate on the potential human rights consequences of California's carbon policies in the Global South. This vocal opposition, which was covered by traditional news media and on social media platforms, slowed down the adoption of a California REDD offset program in the near term. The environmental justice groups' ability to politically pivot and reframe SB 605 also demonstrated their growing power in the capitol. However, this was achieved with mixed outcomes for individual groups and local contexts.

This case of translocal activism, moreover, illustrates how groups with different constituencies can join to take advantage of political openings, as well as underlying common values to propose alternative forms of climate change governance. The translocal coalition developed a conceptual framework that acknowledges complex relationships to place and scale that intersected around race, class, and cultural practices. Through this framework, the coalition challenged powerful actors who often dominate environmental narratives and the policymaking process. The coalition's visibility offers a profound reminder to policymakers of the human scale of carbon markets and also of the purported negative implications that offsets can hold.

These implications apply not just to communities living near California's polluting facilities but also to Indigenous communities in the Global South, whose cultures and livelihoods depend on forests now considered to be tradeable carbon market commodities.[77]

California's Climate in a Global World

> People of color are the new majority in California, and we care about the environment and climate change. These same residents are just beginning to flex their political muscle on climate and environmental issues.
>
> —California Environmental Justice Alliance, *Environmental Justice Scorecard 2014*

At the close of the first day of Governor Schwarzenegger's 2008 Global Climate Summit, I witnessed one of the climate action hero's press aides standing nervously before a crowd of news reporters. The junior aide was tasked with defending the governor from critics who claimed his event was a "shallow Hollywood production." With over 300 attendees from across the globe, it was alleged that the governor had no plan to offset the carbon emissions generated from all that travel. The press aide asserted that emissions would be compensated through the purchase of offset credits, but this plan did not seem to satisfy many in the media. One reporter later quipped that the governor "reportedly purchased carbon credits to offset the burden on the air of his [private jet] travel, but in the end he was a true *Predator* in terms of fuel consumption."[78]

These criticisms were echoed by California environmental justice advocates who viewed the climate summit and Schwarzenegger's global mentality as a new form of "climate colonialism." This form of colonialism was described by advocates as a situation arising when industrialized countries pressure developing nations to conserve forests in ways that benefit only the more powerful, polluting states—an arrangement that exerts control over nature and peoples' livelihoods. In their perspective, while forest conservation is ecologically attractive, it also forms a source of capital for California polluting industries—resulting in what David Harvey calls "accumulation by dispossession."[79]

The anti-REDD movement has gained some power in climate policymaking; it has flexed political muscles and caused rifts among California policymakers. In recent years, California's debate over international linkages and the scope of its climate change regime has grown to include Indigenous and environmental justice activists alongside communities of climate experts. Environmental justice groups are expanding the state's framing of climate change and challenging the spread of California's regulatory programs to the Global South.

Although state regulators remain committed to pursuing forest offsets in Mexico and Brazil, environmental justice groups have raised important questions regarding California's ability to ensure reliable emissions reductions while preventing potential human rights violations abroad. Influential business groups and regulators continue to argue that California should play an important international role in preserving forests and decreasing the cost of mitigating greenhouse gas emissions. The state's pending decision on whether to approve REDD offsets, however, will likely no longer depend solely on economic demand for such offsets or global science documenting the benefits of carbon sequestration. Rather, it will be determined by regulators' ability to build confidence in their capacity to develop effective, politically supported policies that address public health and social equity issues within and beyond California's borders. The notion of community-led solutions is becoming part of a larger framework of climate change policy in California.

Since the passage of SB 605, environmental justice issues and the local effects of globally conceived climate action have taken a central place in debates at the highest levels of policymaking. As countries around the world prepared for the 2015 UN Climate Change Conference in Paris, France, California organized its own preconference event. "Global Climate Negotiations: Lessons from California" brought together government officials, business leaders, scientists, and environmental justice groups. They showcased the state's efforts to meet its overarching goal of reducing global greenhouse gases while providing multiple benefits to the most polluted communities.

Of particular note was the inclusion of the panel "Climate Solutions from the Bottom Up" (fig. 6.5). The panel discussed the man-

Figure 6.5. The panel "Climate Solutions from the Bottom Up" at California's Global Climate Negotiations symposium. *Left to right:* the former California EPA secretary Terry Tamminen; state senator Kevin de León; the IPCC chair, Rajendra K. Pachauri; and the mayor of Long Beach, Bob Foster. Courtesy of the California Air Resources Board

dates in the Climate Change Community Benefits Fund (SB 535) that direct investments in clean energy, transportation, and green jobs to disadvantaged communities, as well as the new measures requiring the board to tackle black carbon. Throughout the panel, Senator Kevin de León (author of SB 535) stressed that UN leaders must learn from the California experience; otherwise, they risk widening climate disparities globally. Echoing these comments, Rajendra K. Pachauri, chair of the Intergovernmental Panel on Climate Change, remarked that "California is being watched by the rest of the world. . . . If California sets this blazing trail, the rest of the world will have something to follow."[80]

Questions of accountability and Indigenous perspectives on international forest-based offsets have also continued to play an important but complex role in public discussion in California. At an April 2016 board hearing in Sacramento, environmental justice advocates reminded regulators of their opposition to a California REDD program in Brazil and Mexico (figs. 6.6 and 6.7). They pointed to purported human rights problems within a forest pre-

Figures 6.6 and 6.7. Vertical Activism: California environmental justice groups and the Nigerian delegation protesting REDD projects at the state air resources board workshop on cap-and-trade, April 28, 2016. Photos by the author

serve in Nigeria (a forest task force member) and brought a Nigerian environmental activist to the hearing to testify on the negative effects of a recent conservation agreement. However, traditional environmentalists and business groups have also found important allies in Indigenous leaders who support international forest offsets. Over the past 17 years, the Yawanawa people of the Rio Gregório Indigenous Territory in Acre, Brazil, have worked with the Aveda cosmetics company. In exchange for preserving forest lands, they receive revenue-sharing opportunities from sales of Aveda products. Tashka Yawanawa, chief of the Yawanawa, spoke in support of a REDD dialogue between governments and Indigenous peoples, which, he argued, could enable stable economies that protect the natural resources his people depend on.

California environmental justice advocates, while supportive of the plight of forest peoples, questioned the applicability of the Yawanawa experience to an implementable California REDD program. The advocates noted that the Yawanawa is a larger tribe that has formal title to its lands and, as a result, stronger political standing to negotiate benefit-sharing agreements with governmental and business entities. As Indigenous voices become more present in these debates, it is clear that they cannot be reduced to a single message. Instead, needs and priorities are shaped by the size and political standing of each group, the legal status of its lands, and a host of other factors that determine outcomes for Indigenous peasant farmers and other forest peoples whose land tenure is disputed and targeted for conservation or development projects.

The translocal opposition to offsets has remained steady in Sacramento as activists' interests merged with those of policymakers seeking greater influence on the state's climate change policies. According to a senior capitol staffer, if the state does implement REDD, it will only be after careful deliberation and the inclusion of social safeguards for the most vulnerable communities. In the staffer's assessment, the environmental justice community has had a real impact on the policy debate: "Ten years ago regulators would have jammed this through without much debate" in their eagerness to spread the state's climate change program around the world. The staffer went on to say that the demographic makeup of the legislature now is very different from when the Global Warming Solutions

Act was adopted in 2006. The legislature is now more racially diverse, and many of the new legislators are more empathetic toward environmental justice concerns.[81]

Richard Corey, the board's executive officer, best exemplified these changing political dynamics when he spoke to Brazilian delegates at a September 2016 annual summit of the forest task force in Jalisco, Mexico: "We think offsets play an important role in the program . . . but we need to go through a continued public process to work through every issue that folks have raised, concerns that they raised."[82] Later that month, at a REDD panel that I hosted at the Yale School of Forestry, the California Air Resources Board member Dean Florez commented that regulators are walking a fine line in balancing in-state political pressures with a goal of combating global climate change: "Are we trying to save California, or are we trying to save the world? In some sense, the answer may be both." Florez identified international forest offsets as one of the most controversial issues facing the board. He added that the translocal coalition had done a good job in highlighting the alleged negative aspects of offsets, which makes it more difficult to quickly approve the program.[83]

This question of whether California can "save the world"—and whether it can protect human rights in the Global South—emerged again in May 2017, during a meeting in the Xapuri region of the state of Acre, Brazil. The gathering brought together Apurina, Huni Kuin, Jaminawa, Manchineri, and Shawadawa Indigenous peoples, representatives of forest communities, and rubber tappers who are potentially affected by offsets projects. The meeting focused on the impacts of climate change policy on their access to territory, their forests, and their livelihoods. Among the organizers of the meeting was Dercy Telles, who had worked alongside the internationally known labor and land rights advocate Chico Mendes. During the meeting, Telles made a video statement for California policymakers and environmental justice advocates. She described how their forest communities view offsets as "nothing but a bunch of false solutions to global warming . . . based on lies and on selling illusions to the less privileged" and told "people in California . . . that we are fighting the same fight in solidarity, . . . and we are going to keep fighting for as long as we have strength and courage."[84]

At the end of the meeting, those attending produced the Xapuri Declaration, which outlined their reasoning for opposing offset projects and associated tropical forest management programs. Following the declaration's publication, signatories were allegedly threatened by state and business actors in Brazil. In response, they contacted board officials in Sacramento to report the intimidation. In emails and phone messages delivered in Portuguese, they inquired what role the board had in moderating or preventing this type of conflict. At the time, the board did not have anyone who spoke Portuguese, nor did it have the political authority to intervene on the Xapuri signatories' behalf.[85]

As a prominent environmental justice advocate argued, this experience illustrates California's limited capacity to "save the world" and protect human rights abroad. Sounding exasperated, the advocate said, "Who at the board speaks Portuguese and can explain to indigenous peoples in Brazil their options? So the board is not a United Nations mediating agency. So how are they going to respond when these grievances, uprisings, or military conflicts happen over their forestry offsets? What is the board going to do when there are cases where local peoples are fighting about a California program?"[86]

The translocal coalition against offsets represents a new model of engagement with climate change and offers alternative paradigms of what forests are and what global environmental protection should be. Through their efforts, anti-REDD activists are challenging the perspectives and scales of analysis and action that are considered valid within California's decision-making on climate change. The translocal coalition posed a fundamental question: Who has the power to protect nature and humanity from the existential threat of climate change? The answer has emerged through various forms of conflict and collaboration. California's climate change policy increasingly depends on the ways in which it incorporates marginalized voices and diverse viewpoints from within the state and around the globe. Although the fate of REDD is still unresolved, the translocal coalition has nonetheless highlighted the complexity of global climate action and confronted mainstream worldviews of environmentalism. In doing so, it is empowering communities globally to imagine new possibilities in the fight against climate change.

CHAPTER SEVEN

California Complexity and Possibility

California is often seen as a homogenous entity that uniformly values environmentalism and climate action. This image universalizes the idea of climate change and detaches it from its cultural settings. It also obscures how the localization of environmental policy and science within the state involves processes of public contestation and legitimation. The complexity of California's environmentalism manifests itself in diverse ways. For example, in the San Joaquin Valley—a conservative region where agriculture and oil interests dominate—contention over air quality regulation is fueled by an antiregulatory climate in Washington, DC, and a coalition of businesses and local air pollution regulators.[1] Seizing upon the 2016 election of Donald Trump as president, the local air pollution chief helped lead a national campaign to eliminate what the coalition saw as the federal Clean Air Act's "costly bureaucratic red tape," including an Obama administration rule that strengthened ozone standards. In response to this campaign, the Trump administration delayed implementation of the new standards.[2]

In another articulation of the state's environmental complexity, on July 17, 2017, the California legislature reauthorized the state's cap-and-trade program in a rare bipartisan effort. Challenging conservative leaders and party activists, eight Republicans, including the assembly Republican leader, joined with Democrats to continue the cap-and-trade program until 2030. Six weeks later, however, facing a torrent of anger from conservative critics, the Republican leader was ousted by his colleagues. Environmental justice activists were similarly angry but for different reasons. The California Environmental Justice Alliance claimed that the reauthorization of cap-and-trade provided too many concessions for industry and lacked the specificity needed to ensure improved health and local air quality. It pointed to the fact that powerful business groups, including the California Chamber of Commerce and associations representing manufacturers, oil companies, and agriculture interests, supported the reauthorization. Among the business concessions included in the reauthorization was language that preempted local regulatory air districts from imposing on oil refineries additional and more stringent mandates for the reduction of greenhouse gas emissions. However, in an attempt to address environmental justice concerns, the legislature also passed a companion bill, AB 617, the Community Air Protection Program. This important measure established the nation's first state-run community-scale air quality monitoring program and one of the world's first programs of this type to detect elevated exposures (pollution hot spots) at the neighborhood scale.[3]

AB 617 also included several other significant provisions to fast-track direct reductions of local pollutants. It accelerated deadlines for existing large stationary emitters (that is, oil refineries and power plants) to install the best available pollution control technology and increased the monetary penalty for exceeding pollution levels, the first penalty increase in 40 years.[4] All in all, the reauthorization of cap-and-trade highlighted the interplay of contrasting economic, environmental, and political interests connected with California's climate change policies. As Alice Kaswan observed, "Compromises abounded, with industry achieving significant concessions in the primary cap-and-trade extension, and environmental justice stakeholders achieving a companion bill that, although not directly addressing concerns about cap-and-trade, nonetheless renews attention on the cumulative burdens many communities continue to endure."[5]

These disparate examples point to a sense of critical complexity and also of possibility—of other ways of imagining climate politics.

They make visible the ways in which conflict and collaboration have broadened California's climate policy agenda. In a sense, they illustrate new imaginaries of climate change. The word "imaginary" is related to similar terms such as "imagination" and "imagery," yet according to Andrew Karvonen, it is "specifically focused on connecting vision and perception and meaning making."[6] A climate change imaginary, thus, is an attempt to develop new relationships among diverse peoples, worldviews, and nature. The notion underscores how environmental justice advocates have gained significant ground in making the improvement of local air quality and direct community benefits a political prerequisite for implementing climate change policies. While environmental justice activists could not overturn the cap-and trade program, their movement has strengthened in recent years. Its members, as state senator Henry Stern pronounced, have "defined the terms of the [climate change] debate even if they don't feel like they won the debate."[7]

These advocates have advanced an aggressive and robust "high stakes experiment" in the crafting of climate policy and advancement of environmental justice values. The movement transformed residents exposed to injustices into organized, politically engaged groups that have debated powerful corporate and state interests over environmental inequities. Advocates also played a pivotal role in pressuring policymakers to consider the health and environmental consequences of their decisions for low-income communities of color.[8] Yet much work remains to ensure that state agencies continue to implement policies and programs equitably. Scholars have documented a long history of instances in which opponents of environmental justice policy have been able to weaken its implementation. This type of move creates processes that require community buy-in without generating significant change. One outcome is that environmental justice groups are prevented from translating their values to regulators, which results in ineffectual programs.[9] As the state's climate programs unfold over the next decade, we will witness how regulators put concepts of environmental justice into practice. Will they institutionalize the definition of environmental justice as advanced by activists? Or will strong ties between powerful state and business actors ensure that the status quo continues? The sociologist Jill Harrison suggests that the institutionalization of environmental justice principles requires "taking seriously inequality, oppression, a lack of participatory parity, and inadequate basic capabilities in all aspects of environmental regulatory practice." From this perspective, environmental regulations must take into account the effects of various forms of injustice to tackle the ways in which they "both deepen environmental problems and render them invisible within current regulatory practice."[10]

The adoption of AB 617, with its focus on local air quality, however, is a significant departure from 2006, when Governor Schwarzenegger signed the Global Warming Solutions Act. At that time, no companion measure existed with the aim of helping disadvantaged communities. Beginning in 2012, California adopted a wave of legislation inserting environmental justice elements into the state's climate change policies. The California Air Resources Board now has an environmental justice–focused senior officer and two board positions dedicated to environmental justice representatives. The state expanded its adaptation strategies to include robust local programs in disadvantaged communities. In addition, thanks to activists' campaigns, nearly 50 percent of the state's cap-and-trade revenue benefits disadvantaged communities—which has resulted in billions of dollars in investments—and half of the carbon offset credits issued (including forest offsets) now must demonstrate benefits within California.[11] As these environmental justice policies are put into practice, the meanings and methods of statutory and regulatory enforcement will likely be contested. Jonathan London et al. argue that understanding the power relations of interested parties in this new policy arena must be a central focus of theory and practice as climate change laws, policies, and programs continue to develop.[12]

Throughout this book, I have examined the culturally contingent nature of climate policy—the assumptions and values that often create conflict between community understandings of local environmental conditions and the prevailing regulatory culture of climate change. Tensions between worldviews, moreover, are often centered on the politics of scale, markets, and race. Activists' advocacy work shows how climate change policy is an ongoing cultural creation that is made and remade through the daily practices of diverse people. Through a reoccurring process of conflict and collaboration, a broad range of individuals and organizations is co-constituting what climate change means. Mike Hulme argues that the tension between worldviews can have a balancing—even creative—impact that yields stronger, more robust approaches to resolving climate change. Furthermore, worldviews are not fixed and can transform over time. Scientific ideas and beliefs about climate change evolve together with the representations, identities, debates, and institutions that give practical effect and meaning to policies.[13] In other words, the ways in which we conceptualize climate change don't just happen. People are behind our government, policies, and environmental values—and they can change their minds.

For instance, at a 2017 UN climate conference in Bonn, Germany, the former California Republican governor Arnold Schwarzenegger made statements that some traditional environmentalists ascribed to the "wrong" climate worldview. He insisted that climate action should now focus more on local air pollutants and their health effects. Schwarzenegger stated, "There aren't enough [climate change] conferences about people who die from cancer because they live too close to the freeway or to a port . . . [or] kids developing asthma and having breathing problems and having to use an inhaler. We're communicating wrong when we always talk about the same thing and throwing around statistics but we never talk about the amount of people that are dying. This is how we can bring people into our crusade and then march forward in a much more successful way than we have."[14]

Schwarzenegger's 2017 position, however, directly contrasted with his worldview when he held public office. As California's governor from 2003 to 2011, he often thwarted efforts to link climate change policy with environmental justice and public health. His changing perspective reflects the broader evolution of California's civic epistemologies of climate change. Through such instances, we can assess the diversification of climate change and politics and trace how scientific facts about the world are fused with social commitments. A local focus on climate change, according to the geographer Mark Pelling, enables us to question the policies and worldviews that establish inequalities and unsustainable relationships with nature in the first place. To resolve the challenge of climate change, critical attention must be placed on the human dimension (that is, on local knowledge, history, and culture) of climate policymaking.[15] And this is precisely how environmental justice groups have targeted their advocacy work in California. They see climate change as an embodied phenomenon and call attention to how the health of their bodies is determined not only by personal choices or genetics but also by a range of social and historical factors that center on race and class.

The environmental justice studies scholar David Pellow further analyzes engagements of the body with socio-ecological threats from states, industries, and other political and economic forces. This approach can provide a critical understanding of how various marginalized populations are treated as expendable. Ryan Holifield contends that their expendability reflects a notion that the health and well-being of communities of color can be "permanently sacrificed for some other interest, whether for the 'common goods' of security or development or simply the private interest of short-term profit."[16] Because of historic racism, Pellow argues, the African American body in particular is subject to continuous forms of disproportionate impacts and erasure in global environmental configurations. In a similar vein, critical race theorists often describe racism as a poison, the presence of which reveals deep contradictions and tensions in the United States, which have resulted in past revolts and wars.[17] In related fashion, Lani Guinier and Gerald Torres propose that "race . . . is like the miner's canary. Miners often carried a canary into the mine alongside them. The canary's more fragile respiratory system would cause it to collapse from noxious gases long before humans were affected, thus alerting the miners to danger. . . . Those who are racially marginalized are like the miner's canary: their distress is the first sign of danger that threatens us all. It is easy enough to think that when we sacrifice this canary, the only harm is to communities of color. Yet others ignore problems that converge around racial minorities at their own peril, for these problems are symptoms warning us that we are all at risk."[18]

Pellow extends the racial metaphor of the miner's canary in spatial terms. He uses the example of communities in the Global South being harmed by international environmental injustices (resulting from hazardous waste dumping, illegal waste trade, or resource extraction by the Global North) and argues that those spaces and the people who occupy them constitute the miner's canary. Such communities are often viewed as disposable by more privileged actors on the global stage. Pellow contends, however, that when social movements organize to "demand that imported toxics be returned to their points of origin, the receiving nations of the Global North serve as a reminder that environmental racism—like racism, class, and gender inequality more generally—threaten us all." Hence the miner's canary signals environmental risk that threatens not only members of socially vulnerable groups but also privileged populations across the globe. Environmental justice activism is often framed in these terms: not only attempting to serve the interests of local communities but also ensuring human well-being at every level.[19]

What Influences the Uptake of Climate Change from the Streets?

Equitable changes to the dominant environmental protection paradigm in California often did not manifest from within regulatory agencies or polluting industries but from an alliance of networked environmental justice advocates.[20] The interconnected case studies in this book are intended to assist scholars, policymakers, and practitioners to work collaboratively with advocates to decrease the divergence between the goals of environmental justice and their implementation in regulatory practice. The California experience provides poignant examples of how, when racialized and undemocratic processes are left unresolved, they can result in contentious climates or ineffective policy. It is undeniable that the numerous substantive changes that environmental justice groups achieved emerged from favorable economic, demographic, and political conditions within the state. Yet California still offers an important example for stakeholders elsewhere of how climate change advocacy and policy can benefit from multiscalar and intersectional approaches. From it, crucial lessons can be drawn about the factors that encourage the adoption of perspectives and action on climate change from the streets.

Engagement with Multiple Forms of Knowledge

Climate change policies are often developed through forums and methods that are least accountable to the people for whom the stakes are highest. This partly results from the assumption that scientific expertise represents the only valid knowledge about climate change. But as I have argued in this book, climate policy is often most successful—politically as well as environmentally—when it emerges from forums where multiple kinds of knowledge are considered. In California, climate change was initially framed as an environmental problem without regard for its human and cultural dimensions. Policy responses emerged from an expert-driven process that emphasized global greenhouse gas reductions as the goal and geographically neutral economic and technological fixes as the solution. In this process, community-based strategies that could integrate climate change interventions with public health outcomes were often excluded. Such approaches privileged experts as the bearers of knowledge about both the problem and the solutions.

Environmental justice groups raised a critical voice, fostered opportunities for constituent education, built coalitions with diverse communities throughout the state, and developed a public record of grievances and deliberations. Through a continual practice of conflict and collaboration, environmental justice communities are no longer merely perceived as impacted; they are now official knowledge producers in climate change governance. They are calling attention to the relevance of the totality of their lived and embodied experience. In doing so, they connect climate policy to historical and current environmental racism, economic inequality, and other sociocultural variables. The movement has influenced policymakers to expand the scope of knowledge and metrics used to address climate change.[21]

For example, in speaking about his negotiations with multiple stakeholders at the 2017 UN climate conference in Bonn, Germany, California state senator Ricardo Lara stated, "So I was having conversations with scientists about alternative metrics that mean something to people. . . . How do we go to Los Angeles and talk about what will happen in asthma clusters, cancer clusters, low birthweight areas? . . . Can we at least start the research to quantify the impacts not just on polar bears, but on the daily lives of people?"[22] Lara's statement represents how embodied knowledge is not an alternative to scientific knowledge; rather, it is something more nuanced, pragmatic, and highly negotiable and collaborative. His engagement with scientists is a reminder that all environmental knowledge is shaped by viewpoints that are partial and situated, and there are limits to all knowledge.[23] Climate change from the streets promotes the development of holistic, interactive, and multiparty processes of knowledge-making to improve environmental health decisions.

Coalitions and Intersectional Activism

Environmental justice groups are often most effective when they can deploy social capital to organize coalitions, develop knowledge frameworks, and build alliances with key policymakers to change the terms of political discourse.[24] Coalition-building is crucial to their ability to act beyond the local scale. Communities with fewer resources and capacity are less likely to persuade officials to implement climate change policies that integrate public health and environmental justice. This is in part because authority in knowledge production is often linked with social status and access to resources. Furthermore, carbon reductionism coincides with political structures that inhibit the capacity for collaborative policymaking across sectors, as well as with budget constraints from competing socioeconomic initiatives.

The formation of the Oakland Climate Action Coalition demonstrates that effective and well-resourced social movements consist of far more than a single nongovernmental organization or group of activists. They form complex networks of organizations mobilized by a collective identity. The Oakland coalition produced a powerful dynamic that generated considerable political and media pressure.[25] Its grassroots narrative was effective in articulating climate change issues, conveying information about tangible, knowable impacts in ways that resonated with diverse constituencies across the city. Motivated by disproportionate environmental burdens, environmental justice groups sought ways to more explicitly link urban planning, public health, and climate change. They advanced community-based projects that reduced global greenhouse gas emissions and envisioned transit-oriented development with affordable housing, renewable energy, local jobs, and urban forestry projects to help cut the risk of asthma and respiratory diseases. With the formation of the coalition, environmental justice groups worked alongside city officials to create a climate action plan that went beyond purely ecological notions of the environment. Members' knowledgeability and the support of politically influential and long-established community stakeholders ultimately persuaded the city to let the coalition facilitate and fund a community-driven climate planning process.

Social movements around climate change, however, are not restricted by scale or locality. Environmental justice groups often scale up local campaigns to increase the likelihood of influencing policymakers and outcomes. Encouraged by their early success in creating an equitable climate action plan in Oakland, members of the coalition joined statewide organizing efforts to amend AB 32 and establish the Climate Change Community Benefits Fund. Environmental justice groups considered the fund an opportunity to link their grassroots work on community-based climate solutions with policies at the state level. Groups such as the Asian Pacific Environmental Network and the Ella Baker Center felt a responsibility to use the lessons learned and their successes at the local scale to influence larger statewide policy. Through the fund, California is now directing hundreds of millions of dollars to disadvantaged communities for climate action planning investments.

Scaling up was also influential in the formation of a translocal climate movement by Indigenous rights leaders and California environmental justice advocates. This movement challenged the spread of California's climate programs to the Global South. While the translocal coalition has not halted the use of international forest offset projects in Mexico and Brazil, its presence provided opportunities for open dialogue about how California can develop an effective and politically supported climate change program that addresses health and equity within and beyond California's borders.

The strength of these coalitions derived in large part from the diversity of their members and their intersectional approach to climate change issues. In Oakland, labor, immigrant, and women's reproductive justice groups joined with faith-based organizations to advance an equitable climate change agenda. Likewise, in 2017, the California Environmental Justice Alliance held town halls throughout the state to unite with other sectors—from immigrant rights to housing rights—to engage in critical movement-building to uphold the rights of all communities of color under the Trump administration. By hosting a series of in-district community town halls with allied organizations such as Causa Justa: Just Cause, Tenants Together, and the United Farm Workers Foundation, activists sought to create a space to educate local elected officials and engage residents in dialogue on important issues affecting them.

Adherence to Carbon Reductionism

Policy and programs based on carbon reductionism often ignore the negative environmental and social impacts of climate change on communities. When climate knowledge remains solely in the realm of experts, communities are disempowered to examine the issue for themselves. Without contextual policies to address equity concerns, climate change policy and action can reinforce current and future health disparities in marginalized communities.

During debates over California's AB 32, there was strong adherence to carbon reductionism in the halls of the capitol and the corridors of the state's bureaucracy. California regulators articulated climate change on a global scale—framing it as an equally distributed environmental problem with no direct health consequences. This framing often separated climate change from the human and local scales, which limited the opportunity for community-based forms of action and analysis. Regulators largely ignored the objections of the state's Environmental Justice Advisory Committee concerning the lack of robust public health analyses of the potential risks and impacts of the cap-and-trade program. The committee alleged that the program would disproportionately harm communities of color and low-income neighborhoods. But since they did not produce analyses to support their claims (due to a lack of capacity and resources), committee members were perceived as being led by emotion more than science. Regulators retained legitimacy and power in climate change governance because they were seen as objective and technically disciplined.[26]

Ethnic and racial divisions also shaped adherence to carbon reductionism. During the drafting of AB 32, environmental justice groups collaborated with legislators of color and found common cause in a shared embodied experience of environmental inequities. By comparison, board regulators charged with implementing AB 32 were mostly white and presumably did not share similar histories of environmental racism. Through the bill's implementation, we see how leadership in climate change policy can be highly homogenous in terms of race and class. When AB 32 moved from the legislature to implementation and the realm of the board's scientific and policy experts, environmental justice groups lost their influence. These variables raise significant questions about how race and class can impact the production of knowledge and solutions around climate change.

Experiments in Climate Change Governance

Climate change from the streets is a form of experimentation. It provides new methods to define climate change as a public concern and to test policy interventions. Through a focus on experimentation, we understand that scientific knowledge, policy decisions, and technical know-how are heterogeneously engineered by a range of people, institutions, and inscriptions. According to Nancy Van House, inscriptions such as policy language, scientific metrics, tools, and maps are central to the process of gaining credibility, as they stabilize experiments and allow them to travel across space and time and combine with other innovations. Access to inscriptions is crucial for environmental justice groups, since they have the potential to both widen participation in the governance of climate change and influence the terms of debate within the policymaking process.[27]

California environmental justice advocates experimented by changing ideas, resources, and relationships as they built coalitions and networks of allies to engage in collaborative problem-solving. Their experimentation with new climate change solutions often began through efforts at the local scale. Initiatives such as Oakland's climate action plan and the Climate Change Community Benefits Fund focused on enacting achievable goals through multiple inscriptions. They developed scientific tools to measure and map cumulative impacts; affordable housing and transit studies; and webinars and surveys linked to advocacy campaigns. These inscriptions helped activists acquire new skills and knowledge as they strengthened relationships with policymakers and experts. As Margaret Keck and Rebecca Abers argue, their experiments were scaled up because others perceived them as valuable. Such recognition can transform the bounds of possibility by making ambitious and formerly impractical ideals seem feasible.[28]

In its early stages, the Oakland climate action plan followed the approach of carbon reductionism. But in the end, the conceptualization of climate change in Oakland was the outcome of extensive experimentation with public participation in advisory processes that brought together multiple forms of knowledge. These processes created new civic epistemologies that required experts to collaborate with a diverse range of stakeholders. As a well-resourced social movement, the coalition gained significant support from policymakers for their environmental justice goals. This helped them transform climate change into an object that is comprehended, measured, and governed on multiple scales.[29]

Environmental justice advocates were subsequently successful in influencing a new civic epistemology of climate change at the state capitol and within regulatory agencies. At first, they adopted oppositional tactics, such as lawsuits to change regulatory behavior and technical practice. The SB 535 Coalition, in contrast, established multiscalar policy approaches to legislatively mandate that environmental justice be at the forefront when implementing climate change interventions. The resulting Climate Change Community Benefits Fund redirected the geographically neutral focus on carbon reductions to new localized measures to address mitigation, adaptation, and health benefits in the most socially vulnerable communities. While many environmental justice groups are philosophically opposed to market-based mechanisms, the SB 535 Coalition embraced pragmatic approaches to ensure that compensation via cap-and-trade revenues is provided first to the communities most impacted by air pollution and poverty.

Through conflict, experimentation, and pragmatic trade-offs, environmental justice groups have revealed that climate change is not simply an environmental problem requiring a singular, top-down policy solution. They have demonstrated the utility of multiscalar approaches that tackle a global environmental problem, as well as public health issues, at the neighborhood level. As Hulme contends, by scaling up community-based approaches and local knowledge, they offer an alternative conceptualization of climate change—not just what it is but what it means in different places and scales, to different peoples, and at different times.[30] In sum, these experimental processes represent a shift of power and authority in environmental governance that traditionally favors elite actors. Climate change from the streets combines technical practice with embodied knowledge to transform existing environmental protection paradigms at multiple scales.

What Is the Next Generation of the Environmental Justice Movement?

> The scale of an actor is not an absolute term but a relative one that varies with the ability to produce, capture, sum up, and interpret information about other places and times.
>
> —BRUNO LATOUR, "Drawing Things Together"

Environmental justice campaigns have been traditionally fought at the neighborhood scale. Many have seen "localism" as endemic in environmental justice activism for various reasons. Those living closest to a pollution source are most at risk from its effects. Shared observations and concerns of neighbors, such as the awareness of cancer clusters or asthma cases, have typically preceded such campaigns. Eric Carter asserts that localism often prevails, moreover, because of the "siege mentality of neighborhoods that have ample grounds to believe they have been targeted" for undesirable land uses.[31]

In California, however, the environmental justice movement is increasingly gravitating toward reciprocal and dialectical relationships between people, place, and scale. Their campaigns are assigning greater attention to the institutional processes that shape environmental health and climate change policy at various scales and locations. Environmental justice actors in the state understand that the movement is situated between the local scales at which the community protests unwanted pollution and the broader geographic scales where problems are produced and can be resolved.[32] California's environmental justice movement is developing campaigns beyond the "parochial" and linking community concerns to regional, statewide, and translocal contexts. However, it is uncertain whether translocal coalitions focused on protest will be sustainable over the coming years and whether they will evolve to develop proactive climate change solutions.

California environmental justice advocates hold a strong understanding that many nations in the Global South face increased risks of drought, wildfires, extreme weather, and disasters from climate change. They further comprehend that these same nations are least responsible for the problem and simultaneously have the lowest capacity and resources to cope with it. For example, in 2013 the Asian Pacific Environmental Network and the California Environmental Justice Alliance supported typhoon relief efforts in the Philippines. They framed their support as an environmental justice effort grounded in shared global injustices: "Low-income communities around the globe are impacted first and worst by climate change. Even though poor communities, like those in the Philippines, have the smallest carbon footprint, these communities directly suffer the consequences from the energy choices made by wealthy nations and corporate polluters. We see the same pattern here in California, where low-income communities are the most impacted by polluting power plants, extreme weather events and face political inaction on their behalf."[33]

Such intersectional perspectives were also a key motivation for California environmental justice advocates to join the 2014 UN Climate Summit and People's Climate March in New York City. At those events and subsequent international climate negotiations, California advocates united with the global voices of Indigenous people, low-income people, and people of color at the frontlines of pollution to demand bold actions to address the crisis. California advocates argued that by fighting the construction of power plants and the state's dependency on fossil fuels, they will help stop the tide of floods, wildfires, rising sea level, and catastrophic storms across the world.[34] However, absent from their arguments was the role that the state of California could directly play in advancing environmental justice projects within the Global South.

Environmental justice leaders currently have no concerted strategy to develop proactive state measures that would help the nations that are most vulnerable to the impacts of climate change and poor air quality. This is concerning because the state seems committed to extend its climate policies globally. State policymakers view California as a member of the global community and envision its cap-and-trade program as part of larger domestic and international carbon markets. A new era of international linkages began when California and the Canadian provinces of Quebec and Ontario fully integrated their cap-and-trade programs in 2017. It may be only a matter of time before the state takes a more aggressive approach in developing climate change policies in Latin America and beyond. These developments raise significant questions about the role that California's environmental justice movement will play as the state's global climate change programs expand. Will the movement produce proactive measures to assist socially vulnerable communities globally, or will it remain a reactive movement philosophically opposed to international carbon market proposals?

California environmental justice advocates have a unique opportunity to collaborate in further experiments that would uplift local perspectives on global climate injustices. One key change could focus on the state's climate policies in relation to global public health issues. For example, the passage of SB 605 expanded the definition of climate change pollutants to include black carbon (see Chapter 6). Throughout the world, black carbon is a major byproduct of diesel engine exhaust from cars and trucks. It is emitted directly into the atmosphere in the form of fine particles and remains there for days to weeks. Black carbon is both a main contributor to climate change and a concern for public health. SB 605 can serve as a model for the Global South and a framework as California seeks to link its climate change programs abroad. The expanding use of diesel fuels and vehicles, along with high rates of urbanization, means that a large proportion of people in Latin America and elsewhere are being exposed to unhealthy levels of black carbon. Thus, organized efforts to control black carbon would have immediate positive public health benefits, as well as contributing to climate change mitigation.[35]

A black carbon strategy would involve targeting centrally fueled urban transport fleets, with a focus on retrofitting the oldest, dirtiest trucks and buses or providing financial incentives to retire and replace them with cleaner, more fuel-efficient models.[36] Strategies such as these could be politically less contentious than implementing forest offset projects on Indigenous lands in the Global South. By proactively defining the terms of the debate around international linkages, environmental justice groups can lead initiatives that have the potential to provide important climate, public health, and other environmental benefits to millions of socially vulnerable people worldwide.

The California environmental justice movement has long proclaimed that if you "care about global climate change, then fight against local air pollution." Although multiscalar strategies have resulted in better safeguards for environmental justice communities within California, the success of a true translocal movement depends on efforts to develop positive and reciprocal justice and health outcomes abroad. Much work still remains to be done to ensure that coupled approaches to climate change and public health are being taken globally. California environmental justice groups can help lead the way.

Toward Alternative Climate Change Futures

California's climate change programs have gained national and international praise. And community residents at the frontlines of pollution deserve significant credit for pushing the state to experiment and imagine alternative climate change futures. Residents from these communities fight every day for solutions to climate change and environmental injustice. The ways in which residents combat climate change are directly informed by their lives in these communities. Their perspectives on pollution are embodied and concerned with its physical, historical, and social dimensions.

From their struggles, we can imagine the multiple ways in which climate change policy can be scientifically sound and socially robust. Through the environmental justice movement's continuous forms of conflict and collaboration, we also understand that climate change solutions are not configured within a single place or scale. Innovative experiments are occurring at every scale, from the streets of Oakland and Richmond to the jungles of Brazil and Mexico. Climate change is thus an object of multiple natures or meanings. It is brought into being and is resolved by ensembles of heterogeneous people, technical practices, and socioeconomic variables. In sum, through the engagement of diverse publics, the scope of climate change is being redrawn to reveal new networks of empowerment and justice. Activists are not only redressing global injustices through experimentation, conflict, and collaboration, they are also imagining and enacting alternative climate change futures.

Afterword

> The right course of action is always a matter of choice, never of fact.
>
> —PAUL DAVIDOFF, "Advocacy and Pluralism in Planning"

ON THE FIRST DAY of the 2018 Global Climate Action Summit in San Francisco, hundreds of protesters blocked summit entrances with their bodies. They chanted and unfolded banners stating "Climate Capitalism Is Killing Our Planet" and "Support Community Solutions." Summit delegates dressed in business attire had to squeeze past them and police to get inside. In front of an international audience, activists were challenging California's status as a progressive bastion of climate action. The summit, organized by California governor Jerry Brown, was intended to serve as a global platform for regions—not nations—to outline plans to reduce carbon emissions. It brought together nearly 4,500 delegates, including mayors, governors, business executives, and celebrities such as Alec Baldwin and Harrison Ford. The scene outside the summit was a symbolic backdrop, a reminder that a powerful mix of conflict and collaboration continues to fuel California's leadership on climate change and environmental justice.

Inside the summit, I witnessed CEOs, governors, and mayors making bold commitments to reduce carbon emissions. But they did not invoke solidarity with the protesters outside. A series of remarks by Brown and the former New York mayor Michael Bloomberg at a press conference held later that morning caught the divisive mood: "We've got environmentalists protesting an environmental conference. It reminds me of people who want to build a wall along the Mexican border to keep people out from a country that we go to for vacations. Something's crazy here," Bloomberg said.[1] It was not until Winnie Byanyima, executive director of Oxfam International, spoke during the open plenary session that I heard explicit references to equity and justice. "Climate change is a political issue, not a technical issue. It's a question of justice and fairness," she said. "It would take my grandfather, who lives in Uganda, 129 years to emit the same carbon as the average American citizen. . . . The emissions that are damaging our planet are being produced by the rich people, but the repercussions are hitting poor people the hardest." Hers was a lone voice among the prominent speakers.

Then an announcement came from the state of California that outlined efforts to move forward with a draft Tropical Forest Offset Standard. The proposed standard (initially published only in English) would provide methods for the state and other subnational governments, such as those of Acre, Brazil, and Chiapas, Mexico, to ensure the accuracy of emissions reductions from future REDD projects—a key step toward implementing the program. Nine philanthropic foundations also committed to contributing $459 million to support forest conservation projects and ensure the recognition of Indigenous peoples' land rights and resource management traditions.[2]

During the summit's lunch break, I returned to the streets. There, US environmental justice and Indigenous leaders from around the world were telling news reporters that the investments and proposed solutions discussed at the summit were not enough. They argued that Indigenous peoples and communities of color are bearing the most severe, immediate impacts of climate-induced disasters and that a more aggressive approach was needed to protect people and address the root cause of the crisis. I had observed similar displays of resistance by 30,000 people during the Rise for Climate, Jobs and Justice March, which had been held a few days prior in downtown San Francisco. Following the march, hundreds of activists also blocked traffic and picketed outside a meeting of the Governor's Climate and Forest Task Force, a group of now 38 subnational governments that California founded in 2008 to promote the adoption of REDD programs worldwide. The protest was organized by Friends of the Earth, Idle No More, It Takes Roots, the Indigenous Environmental Network, and the Climate Justice Alliance.

Outside the task force meeting, protesters burned sage, held prayer rituals, painted street murals, and displayed papier-mâché effigies of Governor Brown to denounce his support of emissions trading systems. Eventually, they negotiated with task force officials to send representatives into the meeting to present a letter demanding that the group be disbanded and that California abandon forest offsets. But task force officials reminded activists that not all Indigenous groups are opposed to the program. Since 2013, the California Yurok Tribe, for example, has generated nearly a million tons of offset credits on its sovereign lands. The tribe has also become an active advocate for similar projects worldwide. Prior to the Global Climate Action Summit, tribal members worked with the Ford Foundation to sponsor a gathering on their territory to help dozens of Indigenous leaders from 15 countries enhance their communication and messaging skills in support of forest offsets. These dynamic engagements, moreover, are a reminder that the settings in which California develops global climate policy and action continue to be viewed through a shifting kaleidoscope of worldviews.[3]

I spent seven days in San Francisco observing the protests, the summit, and the parallel people's convening. I witnessed important and laudable events—the creation of alliances between public and private sectors, the setting of ambitious goals, and a new global leadership determined to honor the Paris climate agreement that President Trump has disavowed. The Global Climate Action Summit, furthermore, played a practical role in underlining that state and local governments do the nitty-gritty work of building infrastructure, setting policy, and reducing emissions.[4]

Perhaps most important was the presence of outspoken and "radical" environmental justice activists to pressure policymakers for change in social, political, and economic structures. At the same time that they are pushing California to alter its thinking, these activists rally against skeptics who claim that the state cannot make a difference. Their activism is a historical reminder that major transformations in society did not happen all at once everywhere or without some form of conflict. Change is often created by a small group of passionate people whose experiments rupture existing practices. They make change from the streets and scale it up elsewhere. This generative mix of conflict and collaboration mobilizes residents and encourages the development of research and data to propose solutions, instead of simply opposing policies. The environmental justice movement has engaged in a sophisticated inside-outside game and is unique in its ability to collaborate with and also work against policymakers, businesses, and regulators.[5] In a time of increased anti-environmentalism, activists perform acts of creative defiance toward not only Washington, DC, but also state and local governments globally. They help spark debate that can blend seemingly incompatible worldviews and foster the integration of ideas and values. Their engagement keeps us honest about the complexities of, and possibilities for, climate action and environmental justice.

Appendix

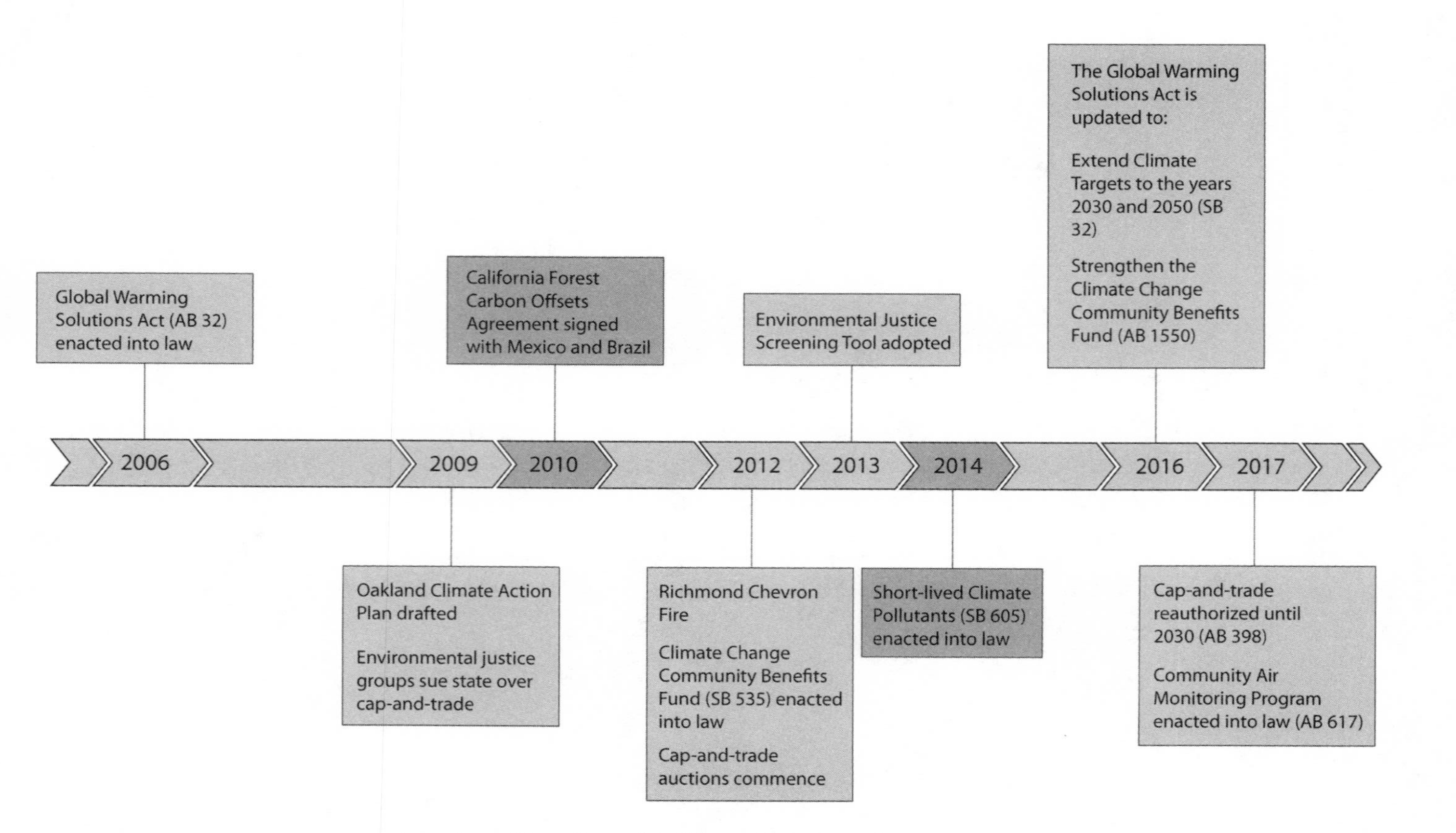

Appendix. 1. Timeline of California's climate change regime

Appendix.2. Map of California

Notes

Chapter 1. Seeing Carbon Reductionism and Climate Change from the Streets

1. Xiang, "Multi-scalar Ethnography," 284. See also Tsing, *Friction*.
2. Xiang, "Multi-scalar Ethnography," 296. See also Çağlar and Glick Schiller, "Multiscalar Perspective"; Williamson, "Multi-scalar Methodology"; Osofsky, "Intersection of Scale"; Barrett, "Necessity of a Multiscalar Analysis."
3. Williamson, "Multi-scalar Methodology," 19. See also Çağlar and Glick Schiller, "Multiscalar Perspective," 2. According to Hari Osofsky, referencing climate change as a multiscalar problem can be a complex and contested concept in both geography and ecology. "Intersection of Scale," 130–33. Geographers define "scale" through four aspects: (1) "a nested hierarchy of bounded spaces of differing size"; (2) "the level of geographic resolution at which a given phenomenon is thought of, acted on or studied"; (3) "the geographical organizer and expression of collective social action"; and (4) "the geographical resolution of contradictory processes of competition and cooperation." Brenner, *New State Spaces*, 9. Ecologists, on the other hand, define "scale" in more technical terms as comprising two parts: grain and extent. "Grain" refers to "the finest level of spatial or temporal resolution available within a given data set," and "extent" refers to "the size of the study area or the duration of the study." Sayre, "Ecological and Geographical Scale," 281.
4. Barrett, "Necessity of a Multiscalar Analysis," 217.
5. Pellow, "Critical Environmental Justice Studies," 223–24; Xiang, "Multiscalar Ethnography," 284. See also Pellow, *What Is Critical Environmental Justice?*; Carmin and Agyeman, *Environmental Inequalities*; Sze, *Noxious New York*; Fischer et al., *Handbook of Critical Policy Studies*; Fischer, *Climate Crisis and the Democratic Prospect*; Hess, Malilay, and Parkinson, "Climate Change"; Curtis and Oven, "Geographies of Health and Climate Change."

6. For human geography, anthropology, and science and technology studies, see Marcus, "Ethnography in/of the World System," 101; Barnes and Dove, *Climate Cultures*, 4–5; Gupta and Ferguson, "Beyond 'Culture'"; Hulme, "Geographical Work," 9; Krauss, "Localizing Climate Change"; Burrell, "Field Site as a Network," 182; Massey, *Power-Geometries*; Graham and Healy, "Relational Concepts"; Latour, *Science in Action*; Haraway, "Situated Knowledges"; Marcus and Fischer, *Anthropology as Cultural Critique*; Urry, *Climate Change and Society*; Giddens, *Politics of Climate Change*. For perspectives from public health and urban planning, see Cummins et al., "Understanding and Representing 'Place,'" 1826–28; Corburn, "Cities, Climate Change"; Corburn, *Toward the Healthy City*, 15.
7. Hulme, "Geographical Work," 9.
8. Marcus, "Ethnography in/of the World System," 105–10.
9. Bulkeley et al., "Climate Justice"; Beck, *World at Risk*. See also Pelling, *Adaptation to Climate Change*. For an in-depth discussion of the historical and broader national and international context in which global regulatory regimes were developed to address climate change, see Fisher, *National Governance*; Norgaard, *Living in Denial*; Klein, *This Changes Everything*; Richter, *Beyond Smoke and Mirrors*.
10. California adopted the nation's first automobile tailpipe emissions standard in 1966, four years before Congress took similar action. After the Middle East oil embargo of 1973, California responded with strong household appliance energy efficiency standards. California's 2009 clean car standard served as a model for the agreement that President Barack Obama forged with automakers in 2011 to double US corporate average fuel economy standards to 55 miles per gallon by 2025. California is a catalyst in the environmental arena in part because of its sheer size and impressive market power. The state's stricter energy efficiency standards have forced automakers and appliance manufacturers to rework their products to maintain access to the nation's largest state economy. Stone, "California Tackles Climate Change"; EDF, "California's Clean Cars Law."
11. California's cap-and-trade regulations limit compliance obligations to entities that emit 25,000 metric tons of carbon dioxide equivalent or more during any of the three years prior to a specified compliance period. CARB, *Annual Report to the Legislature*; Hsia-Kiung, Reyna, and O'Connor, *Carbon Market California*.
12. Zabin et al., *Advancing Equity*, 20. See also United Church of Christ, *Toxic Wastes and Race*; Bullard, *Dumping in Dixie*; Bryant and Mohai, *Incidence of Environmental Hazards*; Cole and Foster, *From the Ground Up*.
13. Mikati et al., "Disparities in Distribution," 480.
14. Cal. Chamber, "Environmental Justice," 133; Sze et al., "Best in Show?," 184; Truong, "Addressing Poverty and Pollution"; Taylor, *Rise of the American Conservation Movement*.
15. Roth, "At Paris Climate Talks." See also Dalrymple, "California Was Ba-

sically Treated"; McDonnell, "Can California Help"; Goodale, "At Paris Climate Talks"; Stockton, "California Marches into Paris."

16. During the conference, the Compact of States and Regions, a consortium of 44 subnational governments from six continents—representing one-eighth of the global economy—announced commitments to cut greenhouse gas emissions by a cumulative 12.4 gigatons by 2030 when compared to business-as-usual projections. More than 400 mayors also signed a compact to measure their cities' emissions and set reduction targets through a climate action plan. C40 Cities, "Cities to Set the Pace."
17. CARB, "Collaboration on Climate Change"; McDonnell, "Can California Help." See also Goodale, "At Paris Climate Talks"; Stockton, "California Marches into Paris."
18. Schaeffer, Pelton, and Kelderman, *Paying Less to Pollute*. See also Naylor, "White House Reverses"; Morris, "Fear of Dying"; Smith, "California Farm Region"; Esty, "Bottom-Up Climate Fix."
19. Dhamoon, "Considerations on Mainstreaming Intersectionality." See also Brah and Phoenix, "Ain't I a Woman?," 76; Clarke and McCall, "Intersectionality and Social Explanation"; Crenshaw, "Demarginalizing the Intersection."
20. Kaijser and Kronsell, "Lens of Intersectionality," 417; Bauer, "Incorporating Intersectionality Theory."
21. Corburn, *Street Science*. This research is also influenced by the work of Agyeman, Bullard, and Evans, *Just Sustainabilities*; Pastor, *State of Resistance*; Pellow, *What Is Critical Environmental Justice?*; Pulido, *Environmentalism and Economic Justice*; Sze, *Noxious New York*.
22. Amin and Thrift, *Cities*. See also Bulkeley, Castán Broto, and Edwards, *Urban Politics of Climate Change*; Pelling, *Adaptation to Climate Change*.
23. IPCC, "Summary for Policymakers," 26.
24. Skocpol, "Naming the Problem," 122. See also Pellow, *Resisting Global Toxics*.
25. Porta and Diani, *Social Movements*. See also Keck and Sikkink, *Activists beyond Borders*. Social movements are often defined as "organizational structures and strategies that may empower oppressed populations to mount effective challenges and resist the more powerful and advantaged elites." Social movements can consist of either formal or informal groupings of individuals or organizations that focus on specific political or social issues. Glasberg, Silfen, and Shannon, *Political Sociology*, 150.
26. Bulkeley, Castán Broto, and Edwards, *Urban Politics of Climate Change*, 18.
27. Curnutte and Testa, "Consuming Genomes," 177–78.
28. In this brief account, I do not seek to provide a complete history of the environmental justice movement. For an in-depth review, see Bullard, *Dumping in Dixie*; Cole and Foster, *From the Ground Up*; Szasz and Meuser, "Environmental Inequalities."
29. Morello-Frosch and Pastor, "Environmental Justice," 256. See also Bullard, *Dumping in Dixie*; Lerner, *Sacrifice Zones*.

30. Pulido, *Environmentalism and Economic Justice*; Harrison, *Pesticide Drift*; Cole and Foster, *From the Ground Up*.
31. Szasz and Meuser, "Environmental Inequalities," 99; Bullard, *Dumping in Dixie*; Zimring, *Clean and White*. See also Morello-Frosch and Pastor, "Environmental Justice."
32. US General Accounting Office, *Siting of Hazardous Waste Landfills*.
33. United Church of Christ, *Toxic Wastes and Race*, 13, 15.
34. Holifield, Chakraborty, and Walker, *Routledge Handbook of Environmental Justice*; Szasz and Meuser, "Environmental Inequalities"; Taylor, *Toxic Communities*. See also Brulle and Pellow, "Environmental Justice."
35. First National People of Color Environmental Leadership Summit, "17 Principles."
36. Cole and Foster, *From the Ground Up*, 32.
37. Omi and Winant, *Racial Formation*, 257. See also Schlosberg, "Reconceiving Environmental Justice."
38. Schlosberg, *Defining Environmental Justice*, 51; Sen, "Capability and Well-Being"; Harrison, *Pesticide Drift*, 15. See also Okereke and Dooley, "Principles of Justice"; Sze and London, "Environmental Justice," 1334.
39. Cole and Foster, *From the Ground Up*; London, Sze, and Liévanos, "Problems, Promise."
40. Pastor, "Environmental Justice," 4. See also Pastor, *State of Resistance*.
41. The initiative's diverse membership included environmental justice, religious, policy, and advocacy groups from hundreds of communities in the United States and globally. Together they developed 10 principles of climate justice, focused on reducing emissions and use of fossil fuels, protecting vulnerable communities, transitioning to renewable energy, community participation in policymaking, and demanding that the United States take a leadership role on climate change. Schlosberg and Collins, "Environmental to Climate Justice," 362–63.
42. Schlosberg and Collins, "Environmental to Climate Justice," 362.
43. Angotti and Sze, "Environmental Justice Praxis," 26–27.
44. Jasanoff, *Designs on Nature*, 255.
45. Jasanoff builds from epistemology studies focused on the nature of knowledge, justification, and the rationality of belief. These studies address such questions as "What makes justified beliefs justified?"; "What does it mean to say that we know something?"; and, fundamentally, "How do we know what we know?" Wenning, "Scientific Epistemology," 3; Jasanoff, *Designs on Nature*.
46. Jasanoff, "Cosmopolitan Knowledge." Civic epistemology also includes a range of knowledge-production processes, including scientific peer review, public participation mechanisms, methods of reasoning, government statistics, standards of evidence, and norms of expertise that typify public debates and political institutions. Miller, "New Civic Epistemologies," 406; Iles, "Identifying Environmental Health Risks."

47. Barrett, "Necessity of a Multiscalar Analysis," 219; Kurtz, "Scale Frames"; Towers, "Political Geography of Scale," 24–25.
48. Iles, "Identifying Environmental Health Risks."
49. Jasanoff, *Designs on Nature*, 255.
50. Lind, "Five Worldviews." See also Palmer, *Cultural Linguistics*.
51. Bentham, *Introduction*. For more on utilitarianism, see Schofield, *Utility and Democracy*; Skorupski, *John Stuart Mill*; Long, "'Utility'"; Habibi, *John Stuart Mill*; Scarre, *Utilitarianism*. David Harvey describes utilitarianism as a conventional paradigm of environmental governance. He acknowledges that while it has achieved environmental benefits, its central concern is "how best to manage the environment for capital accumulation, economic efficiency, and growth." *Justice, Nature*, 375.
52. Cushing et al., "Carbon Trading"; Stavins, "Environmental Justice Lawsuit"; Boyce and Pastor, "Clearing the Air."
53. Kaswan, "Energy, Governance," 492; Schlosberg and Collins, "Environmental to Climate Justice," 364. See also Okereke and Dooley, "Principles of Justice."
54. Jasanoff, "Ordering Knowledge, Ordering Society," 38–40; London et al., "Racing Climate Change," 798; Fischer, *Citizens, Experts*; Hulme, "Geographical Work."
55. The Kyoto Protocol is an international agreement linked to the UN Framework Convention on Climate Change, which commits its parties by setting internationally binding reduction targets for greenhouse gas emissions. It was adopted in Kyoto, Japan, on December 11, 1997, and implemented on February 16, 2005. UN Framework Convention on Climate Change, "Kyoto Protocol." The seven greenhouse gas emissions listed in the Global Warming Solutions Act are carbon dioxide, methane, nitrous oxide, hydrofluorocarbons, perfluorocarbons, sulfur hexafluoride, and nitrogen trifluoride. The Kyoto Protocol does not address nitrogen trifluoride, which was not widely used at the time that it was negotiated. Recent studies have found that this chemical, which is used in the manufacture of consumer items including photovoltaic solar panels, LCD television screens, and microprocessors, has a higher global warming potential than CO_2 and remains in the atmosphere longer. California Health and Safety Code § 38505(g).
56. Exposure to greenhouse gas emissions has effects on human health in concentrated form, for example, in the workplace. Wisconsin Department of Health Services, "Chemical Fact Sheet." However, outdoor exposure levels are considered negligible. CARB, *Climate Change Scoping Plan*, 86.
57. For example, according to the Intergovernmental Panel on Climate Change, changing precipitation or melting snow and ice are altering hydrological systems in many climates and affecting water resources in terms of quantity and quality. Many terrestrial, freshwater, and marine species have

shifted their geographic ranges, seasonal activities, migration patterns, abundances, and species interactions in response to ongoing climate change. IPCC, "Summary for Policymakers."

58. Boyce and Pastor, "Clearing the Air"; Stavins, "Environmental Justice Lawsuit"; Cushing et al., "Carbon Trading."
59. Union of Concerned Scientists, "Why Does CO_2 Get Most of the Attention." References to greenhouse gas emissions quantities in climate policy follow the international convention of using metric tons of CO_2 or CO_2 equivalent when referring to other gases. CARB, *Climate Change Scoping Plan*. A metric ton equals 2,205 pounds.
60. A "carbon footprint," for example, is defined as "a measure of the total amount of [greenhouse gas] emissions of a defined population, system or activity, considering all relevant sources, sinks and storage within the spatial and temporal boundary of the population, system or activity of interest. Calculated as carbon dioxide equivalent (CO_2e) using the relevant 100-year global warming potential (GWP100)." Wright, Kemp, and Williams, "'Carbon Footprinting.'"
61. Hull, *Infinite Nature*, 4–5.
62. Stavins, "Environmental Justice Lawsuit"; Boyce and Pastor, "Clearing the Air"; Cushing et al., "Carbon Trading."
63. Somers and Block, "Poverty to Perversity," 260–61. See also Klein, *This Changes Everything*. The term "market fundamentalism" was popularized by businessman and philanthropist George Soros, who writes, "This idea was called laissez faire in the nineteenth century. . . . I have found a better name for it: market fundamentalism." *Crisis of Global Capitalism*, xii.
64. Wysham, "Carbon Market Fundamentalism"; Dayaneni, "Carbon Fundamentalism."
65. Baptist, "What the *Economist* Doesn't Get."
66. Allenby and Sarewitz, *Techno-human Condition*, 124.
67. CARB, *First Update*, ES1. These greenhouse gas reduction goals were updated in 2016 (see the timeline in the Appendix).
68. Jasanoff, *Fifth Branch*; Fischer, *Citizens, Experts*; Allenby and Sarewitz, *Techno-human Condition*; Ezrahi, *Descent of Icarus*.
69. California Health and Safety Code §§ 38505(d), 38560, 38561.
70. CARB, *First Update*, 104.
71. Peace and Stavins, *Meaningful and Cost Effective*, 1.
72. Some of the state's other complementary mitigation measures include renewable portfolio standards for electric utilities, low-carbon fuel standards, and vehicle fuel efficiency requirements. Nachbaur and Roberts, *Evaluating the Policy Trade-Offs*.
73. CARB, *Climate Change Scoping Plan*; Mazmanian, Jurewitz, and Nelson, "California's Climate Change Policy."
74. Park, *Everybody's Movement*; ICLEI, *Mitigation-Adaptation Connection*.
75. Park, *Everybody's Movement*; Hulme and Mahony, "Climate Change."

76. Barugh and Glass, "Environment, Climate Change," 3.
77. Agyeman, Bullard, and Evans, *Just Sustainabilities*.
78. Boyce and Pastor, "Clearing the Air," 803; Cushing et al., "Carbon Trading." The Intergovernmental Panel on Climate Change has indicated that greenhouse gas mitigation can result in health benefits from reduced air pollution, which can offset the costs of mitigation. Smith et al., "Human Health." A survey of 37 peer-reviewed international studies also found a mean co-benefit of $49 per ton of carbon dioxide avoided. Nemet, Holloway, and Meier, "Incorporating Air-Quality Co-benefits." In a study of emissions reductions in the United States, researchers estimated that the monetized human health benefits associated with air quality improvements have the potential to offset between 26 percent and 1,050 percent of the cost of various carbon policies. Thompson et al., "Air Quality Co-benefits." In other words, each dollar spent on air quality improvements can yield between 26 cents and $10.50 in savings associated with health costs. Mendez, "Local Climate Action Plans." See also Li, "Including the Feedback"; Pittel and Rübbelke, "Climate Policy and Ancillary Benefits"; Burtraw and Toman, "Estimating the Ancillary Benefits."
79. Krieger, *Embodying Inequality*, 2; Krieger, "Embodiment."
80. Jasanoff, "Cosmopolitan Knowledge"; Corburn, *Street Science*.
81. Braveman, "Monitoring Equity," 181.
82. Pellow, *What Is Critical Environmental Justice?*; Des Chene, "Locating the Past."
83. Locatelli, *Synergies*, 1; Stehr and Storch, "Introduction to Papers"; Few, Brown, and Tompkins, "Public Participation."
84. Brace and Geoghegan, "Human Geographies," 287; Iles, "Identifying Environmental Health Risks"; Clark and Murdoch, "Local Knowledge."
85. Fortun, "Multi-sited Ethnography," 73–87.
86. The state enacts hundreds of laws and policies related to the environment each year. The case studies in this book represent key historic moments in California's climate governance.
87. Cummins et al., "Understanding and Representing 'Place,'" 1827.

Chapter 2. Climate Embodiment

1. Quoted in Mulady, "Richmond Refinery Fire."
2. Kay and Katz, "Pollution, Poverty"; Bullard, "Environmental Justice." See also OEHHA, *Environmental Health Screening Tool*; Morello-Frosch and Pastor, "Environmental Justice"; Clark, Millet, and Marshall, "Transportation-Related Air Pollution."
3. Krieger, *Embodying Inequality*, 2; Krieger, "Embodiment."
4. Shah, *Laotian Daughters*.
5. Mulady, "Richmond Refinery Fire"; Kay and Katz, "Pollution, Poverty." See also CBE, "Richmond."

6. Juhasz, "Chevron's Refinery, Richmond's Peril"; CBE, "Richmond"; Kay and Katz, "Pollution, Poverty"; Cushing et al., "Carbon Trading"; Boyce and Pastor, "Clearing the Air"; Pastor et al., "Risky Business."
7. Kaswan, "Climate Change," 27; Boyce and Pastor, "Clearing the Air."
8. Lemos and Morehouse, "Co-production of Science and Policy," 57.
9. Sasser, "Population, Climate Change," 58–61. See also Whitmee et al., "Safeguarding Human Health."
10. Cama, "Report Faults Chevron Safety."
11. Taruc, interview, May 2013.
12. Sasser, "Population, Climate Change," 61; Haraway, "Situated Knowledges."
13. Taruc, interview, August 2017.
14. Berton and Fagan, "Refinery Warning Worked." During the 1999 fire, emergency calls were conducted only in English. "Shelter in place" safety instructions did not reach individuals who spoke other languages common in the region, such as Laotian and Spanish. Taruc, interview, August 2017.
15. For further discussion on how California's initiatives on cumulative impacts preceded work by the federal government, see Liévanos, "Retooling CalEnviroScreen."
16. OEHHA, "Call for Applications," 1.
17. Environmental justice advocate, interview, August 2017.
18. Tickner and Geiser, "Precautionary Principle," 801.
19. Tickner and Geiser, "Precautionary Principle," 801–2; Morello-Frosch and Pastor, "Environmental Justice," 254.
20. Bukowski, "Review of the CalEPA Report," 1.
21. *Inside Cal/EPA*, "Activists, Industry at Odds."
22. Taruc, interview, August 2017.
23. Discussed further in Chapter 3, their concerns also centered on how the state's greenhouse gas reduction programs could induce businesses such as oil refineries to make mitigation decisions resulting in higher emissions of co-pollutants, while greenhouse emissions would decrease on a statewide basis.
24. Kaswan, "Climate Change," 22.
25. EJAC, "Recommendations," December 2008, 8. See also Kaswan, "Climate Change," 22.
26. Taruc, interview, August 2017.
27. Carbon Leadership Forum, "Embodied Carbon Network." Embodied carbon is also referred to as a "carbon footprint," which is an assessment of the amount of carbon dioxide and other carbon compounds emitted due to the consumption of fossil fuels by a particular person, group, etc.
28. Circular Ecology, "Embodied Energy."
29. King, *States of Disease*; Krieger, "Theories for Social Epidemiology"; Pellow, *What Is Critical Environmental Justice?*

30. Kay and Katz, "Pollution, Poverty"; Krieger, *Embodying Inequality*, 3–5.
31. Krieger, "Embodiment and Ecosocial Theory," 109–10.
32. Krieger, *Embodying Inequality*, 5.
33. Krieger, "Embodiment and Ecosocial Theory," 109–10.
34. Krieger, "Embodiment and Ecosocial Theory," 110; Krieger, *Embodying Inequality*, 3.
35. Moore, *To Place Our Deeds*.
36. Quoted in Kay and Katz, "Pollution, Poverty."
37. ODPHP, "Social Determinants of Health"; King, *States of Disease*, 44; Braveman et al., *Housing and Health*; Braveman and Egerter, *Overcoming Obstacles to Health*.
38. King, *States of Disease*, 46.
39. King, *States of Disease*, 46–47; ODPHP, "Social Determinants of Health."
40. Scheper-Hughes and Lock, "Mindful Body," 7–8.
41. Mendez, "Local Climate Action Plans"; Shonkoff et al., "Climate Gap."
42. Maibach et al., "Reframing Climate Change." See also Lemos and Morehouse, "Co-production of Science and Policy"; Hulme, "Problems with Making and Governing"; Cohen and Ottinger, "Environmental Justice"; London et al., "Racing Climate Change"; Fischer et al., *Handbook of Critical Policy Studies*; Fischer, *Climate Crisis and the Democratic Prospect*.
43. Carmin and Dodman, "Engaging Science," 229.
44. Carmin and Dodman, "Engaging Science," 223.
45. Cartwright et al., *Climate Change*, 1–3; Hallegatte, Lecocq, and de Perthuis, *Designing Climate Change*.
46. Cartwright et al., *Climate Change*, 3.
47. Rosenthal and Brechwald, "Climate Adaptive Planning," 220; Young, *Inclusion and Democracy*; Burayidi, *Urban Planning*. See also Umemoto, "Walking in Another's Shoes."
48. Innes, "Planning Theory's Emerging Paradigm," 183. See also Forester, *Planning in the Face of Power*.
49. Jasanoff and Martello, "Knowledge and Governance," 347.
50. Cohen and Ottinger, "Environmental Justice," 3–5.
51. Ezrahi, *Descent of Icarus*, 69. See also Turner, "What Is the Problem"; Cohen and Ottinger, "Environmental Justice," 3–5.
52. Niemeier et al., "Characterizing the Impacts," 131. See also Corburn, *Street Science*.
53. Corburn, *Street Science*, 50. "Local knowledge" is often defined in the international development literature as (1) information linked to a specific place, culture, or identify group; (2) dynamic and evolving knowledge; (3) know-how belonging to groups of people who are intimate with the natural and human system within which they live; and (4) knowledge that has some qualities that distinguish it from formal scientific knowledge. Adapted from Corburn, *Street Science*, 50.
54. Corburn, *Street Science*, 50; Corburn, "Community Knowledge," 159.

55. Jasanoff, *Fifth Branch*, 76–77.
56. Liévanos et al., "Uneven Transformations," 204–6; Jasanoff, *Fifth Branch*; Knox-Hayes, *Cultures of Markets*, 15.
57. Brown et al., "Embodied Health Movements," 50–55. See also Brown, *Toxic Exposures*.
58. Morgen, *Into Our Own Hands*, 230.
59. Corburn, "Community Knowledge," 152; Iles, "Identifying Environmental Health Risks"; Cole and Farrell, "Structural Racism."
60. González et al., "Participatory Action Research," 89; Minkler and Wallerstein, *Community-Based Participatory Research*, 7. See also Angotti and Sze, "Environmental Justice Praxis."
61. McAdam, McCarthy, and Zald, "Opportunities, Mobilizing Structures"; Corburn, "Community Knowledge."
62. Briggs, "Use of Indigenous Knowledge," 107–8; Schroeder, "Geographies of Environmental Intervention."
63. Cleaver, "Paradoxes of Participation," 605.
64. Corburn, *Street Science*, 202.
65. Farrell, "Just Transition"; Scheper-Hughes and Lock, "Mindful Body," 6–14.
66. Kaswan, "Climate Change," 41.

Chapter 3. Contentious Capitol Climates

1. Montañez parlayed her leadership skills and work ethic to be appointed by the assembly speaker as chairperson of the Assembly Rules Committee, becoming the first woman to hold that influential position. Montañez, interviews, April 2006 and December 2017.
2. EJ Matters, "To Address Climate Change."
3. Jessop, "Liberalism, Neoliberalism," 463; Holifield, "Neoliberalism and Environmental Justice Policy," 203–4.
4. Liverman and Vilas, "Neoliberalism and the Environment," 329. See also Harvey, *New Imperialism*; Lansing et al., review symposium, 2391; Lohmann, "Performative Equations," 170–73.
5. Schlosberg, "Reconceiving Environmental Justice." AB 32 committed the state to reduce its greenhouse gas emissions to 1990 levels by 2020—a reduction of approximately 15 percent below emissions expected under a "business as usual" scenario. California Health and Safety Code § 38565.
6. Environmental justice advocate, interview, May 2013; London et al., "Racing Climate Change"; CARB, "AB 32 Scoping Plan Functional Equivalent Document."
7. Castro, "Latinos and the Environmental Movement"; traditional environmental organization member, interviews, May 2013 and August 2017; senior capitol staff member, interview, May 2013.
8. Traditional environmental organization member, interview, August 2017.

While the Asian Pacific Islander and African American legislative caucuses had a presence in the capitol at the time, their membership was much smaller. In the Natural Resources Defense Council's AB 32 strategy documents, only the Latino Legislative Caucus was listed among the four influential legislative voting blocs to strategically lobby. The other voting groups included "Business Democrats" (which had significant overlap with Latino members), "Moderate Republicans," and "Strong Environmental Supporters." NRDC and E2, "Making a Difference."

9. Senior capitol staff member, interview, May 2013; former Latino Caucus member, interview, August 2013; traditional environmental organization member, interview, May 2013. See also Adler, "New Calif. Legislature"; Bernstein, "California Democrats."
10. Former Latino Caucus member, interview, August 2013.
11. Environmental justice advocate, interview, May 2013.
12. Senior capitol staff member, interview, May 2013; traditional environmental organization member, interviews, May 2013 and August 2017.
13. Latino Caucus member, interview, August 2013.
14. Cal. Chamber, "Job Killers."
15. Latino Caucus member, interview, August 2013; traditional environmental organization member, interviews, May 2013 and August 2017.
16. Latino Caucus member, interview, August 2017; Sze et al., "Best in Show?," 182; environmental justice advocate, interview, May 2013.
17. Quoted in Park, *Everybody's Movement*, 19.
18. EDF, "'Getting to Yes,'" 4.
19. Traditional environmental organization member, interviews, May 2013 and August 2017. A former Latino Caucus member also expressed similar views. Interview, August 2013.
20. EDF, "How Cap and Trade Works."
21. Kaswan, "Climate Change," 14–15; Cushing et al., "Carbon Trading." AB 32 defines "allowance" as "an authorization to emit, during a specified year, up to one ton of carbon dioxide equivalent." California Health and Safety Code § 38505(a).
22. Quoted in EJ Matters, "Environmental Justice Advocates."
23. Nachbaur and Roberts, *Evaluating the Policy Trade-Offs*.
24. Lejano and Hirose, "Testing the Assumptions"; Drury et al., "Pollution Trading." See also Boyce and Pastor, "Clearing the Air"; Cushing et al., "Carbon Trading."
25. Senior capitol staff member, interview, May 2013; Rabe, "Governing the Climate"; OECD, "Command-and-Control Policy." "Command" refers to government-mandated standards for emissions that regulated entities (polluters) must follow. "Control" is the manner in which standards must be met (i.e., installation of pollution control technologies) to avoid sanctions.
26. Sze et al., "Best in Show?," 182; Boyce, memorandum; California Health and Safety Code §§ 38570, 38590, 38591.

27. Latino Caucus member, interview, August 2013.
28. Traditional environmental organization member, interview, August 2017.
29. Former Latino Caucus member, interview, August 2013; traditional environmental organization member, interview, August 2017.
30. Cal. Chamber, "Job Killers"; Martin, "Núñez Slams Governor"; traditional environmental organization member, interview, August 2013.
31. Roland-Holst, "Economic Growth"; Martin, "Núñez Slams Governor"; California Legislative Information, unofficial ballot on AB 32; traditional environmental organization member, interview, May 2013.
32. California Channel, "Assembly Floor Vote on AB 32."
33. Holifield, "Neoliberalism and Environmental Justice Policy," 203–4; Jessop, "Liberalism, Neoliberalism."
34. Holifield, "Neoliberalism and Environmental Justice Policy." According to Holifield, the US EPA's environmental justice strategy was developed in response to President Bill Clinton's 1994 Executive Order 12898, "Federal Actions to Address Environmental Justice in Minority Populations and Low-Income Populations." "Neoliberalism and Environmental Justice," 203–4.
35. California Channel archives, "Assembly Floor Vote on AB 32."
36. Núñez, "Connecting the Dots."
37. Rabe, "Governing the Climate," 51–52.
38. Schwarzenegger, "Executive Order S-20-06." AB 32 defines "cost-effective" in terms of the "cost per unit of reduced emissions of greenhouse gases adjusted for its global warming potential." California Health and Safety Code § 38505(d). While the California Air Resources Board is considered an independent body, the governor proposes its annual budget, appoints its members, and selects its chair (subject to state senate confirmation) and is consulted in the hiring and termination of executive staff.
39. Senior Schwarzenegger appointee, interview, August 2013; environmental justice advocate, interview, May 2013; environmental justice advocate, interview, September 2016. See also London et al., "Racing Climate Change."
40. Quoted in O'Brien, "Schwarzenegger Misled on Global Warming"; traditional environmental organization member, interview, August 2017.
41. Quoted in Rosenberg, "Gov. Schwarzenegger's Global Warming Act." In 2005, the governor also attempted to appoint Cindy Tuck, a lobbyist for industry, as the chair of the board. Her selection was widely opposed by Democrats and environmentalists. The California State Senate, in its constitutional capacity of "advise and consent," rejected her within a month of the appointment. *Capitol Weekly*, "Former Air Board Nominee."
42. Quoted in Harmon, "Schwarzenegger Pressured Air-Quality Board"; Martin, "Núñez Slams Governor."
43. Rosenberg, "Gov. Schwarzenegger's Global Warming Act"; *Capitol Weekly*, "Former Air Board Nominee."

44. Environmental justice advocate, interview, June 2012; Lerza, *Perfect Storm*; Sze et al., "Best in Show?"
45. Williams, "Testimony for the Joint Hearing."
46. Quoted in Sze et al., "Best in Show?," 182.
47. Lerza, *Perfect Storm*; EJAC, "Recommendations," December 2008; Andrew, Kaidonis, and Andrew, "Carbon Tax." The carbon tax is itself a market-based mechanism. However, the committee recognized that in "the neoliberal political climate (personified by famously business-friendly Governor Schwarzenegger) a carbon tax was as close as they could come to an alignment with their values." London et al., "Racing Climate Change," 795.
48. Hecht, "Reflections on Environmental Justice." See also EJAC, "Recommendations," December 2008.
49. Grimes, "Anti-democracy Bill." The committee also opposed the proposed cap-and-trade system because it lacked strong emissions standards and monitoring programs. Members accused the board of resisting the adoption of such rules despite their use in the US EPA's acid rain trading program—generally the program that cap-and-trade proponents most frequently identify as successful. US EPA standards required all polluting facilities to install "Continuous Emissions Monitors" to verify emissions levels and ensure compliance. Yet the board had not proposed such monitoring in its cap-and-trade proposal. The committee claimed that without underlying emissions standards and monitoring, establishing correct allocation levels for pollution credits would be impossible. Overallocation, they argued, would strip away the incentive for businesses to reduce emissions because emissions credits remain cheaper than adapting cleaner technologies. EJAC, "Recommendations," October 2008.
50. EJ Matters, "Declaration."
51. EJAC, "Recommendations," October 2008. See also Cole and Farrell, "Structural Racism," 276.
52. Cole and Farrell, "Structural Racism," 278; Rabin, "Expulsive Zoning," 103–6.
53. Cole and Farrell, "Structural Racism," 278.
54. Shonkoff et al., "Minding the Climate Gap," 174.
55. Cushing et al., "Carbon Trading"; Pastor et al., "Risky Business," 76–77; EJAC, "Recommendations," December 2008.
56. Market Advisory Committee, *Recommendations*.
57. Quoted in Schwebs, "Global Warming I."
58. Bailey et al., "Boosting the Benefits," 3; CARB, "Appendix H," H-13; traditional environmental organization member, interview, August 2017.
59. EJAC, "Recommendations," December 2008; Ezrahi, *Descent of Icarus*, 69; environmental justice advocate, interview, June 2012; London et al., "Racing Climate Change."

60. Faber and McCarthy, *Green of Another Color*, 31; Park, *Everybody's Movement*, 37.
61. EJAC, "Recommendations," December 2008.
62. Former Latino Caucus member, interview, August 2013.
63. Osofsky, "Intersection of Scale," 130–33.
64. Jonas, "Scale Politics of Spatiality," 258.
65. Scott, *Seeing like a State*; Towers, "Political Geography of Scale," 31–32.
66. Quoted in Ostrander, "California's Cap and Trade"; Lebel, Garden, and Imamura, "Politics of Scale," 1.
67. Towers, "Political Geography of Scale," 24; Kurtz, "Scale Frames," 888; Williams, "Environmental Injustice."
68. Kurtz, "Scale Frames," 894; Towers, "Political Geography of Scale," 31–32. See also Brenner, "State Territorial Restructuring."
69. CARB, *Climate Change Scoping Plan*; Doerr, "Smart Businesses." Though never explicitly stated, another factor influencing the board's preference for cap-and-trade was the politically untenable nature of a carbon tax; in California, adoption of a tax requires a two-thirds supermajority vote in the legislature. Under cap-and-trade, a "fee" (emissions allowances sold at auction) is set each quarter by the board, allowing elected officials to avoid voting for a tax. Political dynamics in the capitol at the time, which continue today with the addition of more politically moderate members, made passage of a tax on industry difficult. Senior capitol staff member, interview, August 2017.
70. Martinez-Alier, Munda, and O'Neill, "Weak Comparability of Values"; London et al., "Racing Climate Change," 792.
71. De la Cruz, "'Cap and Trade Will Fail.'" See also Association of Irritated Residents v. California Air Resources Board, No. CPF-09-509562 (CA Super. Ct., SF County, 2011); Egelko, "Calif. Cap-and-Trade Plan."
72. Association of Irritated Residents v. California Air Resources Board, No. CPF-09-509562, slip. op. (CA Super. Ct., SF Cty., March 18, 2011).
73. Stavins, "Environmental Justice Lawsuit."
74. Sadd et al., "Playing It Safe."
75. Quoted in Rogers, "California's Global Warming Law." Núñez left political office on November 30, 2008.
76. CARB, "AB 32 Scoping Plan Functional Equivalent Document"; Parino et al., "Comments," 9.
77. Quoted in Cart, "California Becomes First State."
78. CARB, "Adaptive Management"; National Research Council, *Adaptive Management*.
79. De la Cruz, "'Cap and Trade Will Fail.'"
80. Núñez, "Cap-and-Trade." See also Hsia-Kiung, Reyna, and O'Connor, *Carbon Market California*; traditional environmental organization member, interview, August 2017.
81. Hulme, "Problems with Making and Governing."

82. Park, *Everybody's Movement.*
83. Liévanos, "Certainty, Fairness."
84. Former Latino Caucus member, interview, October 2017.
85. Environmental justice advocate, interview, August 2017.
86. Environmental justice advocate, interview, August 2017.
87. DeBacker et al., *Pathways to Resilience*, 53–54.

Chapter 4. Changing the Climate from the Streets of Oakland

1. Material in this chapter has been adapted from Mendez, "From the Street."
2. Ransby, *Ella Baker*; Kirsch, interview, October 2011.
3. Mendez, "Local Climate Action Plans"; Rosenthal and Brechwald, "Climate Adaptive Planning."
4. Hulme, "Problems with Making and Governing"; CARB, *Climate Change Scoping Plan*; Moser and Dilling, *Climate for Change*; Irwin, *Citizen Science*.
5. Brace and Geoghegan, "Human Geographies," 287.
6. Jasanoff, *Designs on Nature*, 255. I selected the city of Oakland based on research and analysis conducted for a peer-reviewed article that surveyed municipal climate action plans in environmental justice communities. Through interviews with urban planners and a content analysis of local plans, I assessed how California cities with high levels of pollution and social vulnerability address climate change and public health. The analysis showed how Oakland is a leader on community-based climate planning approaches. Mendez, "Local Climate Action Plans."
7. Environmental justice advocate, interview, August 2013.
8. Broad, *More than Just Food*; Bulkeley et al., "Climate Justice," 918.
9. Garzón et al., *Community-Based Climate Adaptation Planning*; Minkler and Wallerstein, *Community-Based Participatory Research*.
10. CalEnviroScreen uses existing environmental health and socioeconomic data to create a "cumulative impacts" score for communities across the state. The tool compares areas of the state with one another, creating a relative ranking. An area with a high score would be expected to experience greater cumulative impacts as compared to areas with low scores. Faust et al., *CalEnviroScreen2.0*.
11. Bulkeley et al., "Climate Justice." See also Douglas et al., "Coastal Flooding"; Rosenthal and Brechwald, "Climate Adaptive Planning"; Rosenzweig et al., "Cities Lead the Way"; Corburn, "Cities, Climate Change."
12. Kirsch, interview, October 2011.
13. OCAC member, interview, May 2013. Life-cycle analysis is similar in scope to an "embodied carbon" analysis (see Chapter 2).
14. Most cities participate in the Cities and Climate Protection Program developed by ICLEI—Local Governments for Sustainability (founded in

1990 as the International Council for Local Environmental Initiatives). The program helps cities establish targets for carbon emissions reductions by developing greenhouse gas inventories and mitigation strategies, including provisions for monitoring and evaluation. Alber and Kern, "Governing Climate Change."

15. Kirsch, interview, October 2011.
16. OCAC member, interview, May 2011.
17. CARB, *Climate Change Scoping Plan.*
18. Quoted in Ella Baker Center and OCAC, *Toolkit.*
19. OCAC Steering Committee member, interview, May 2013.
20. OCAC Steering Committee member, May 2013. After the city released its first and second drafts, the coalition submitted edits using tracked changes, which allowed the city to copy and paste their suggestions into the updated and final drafts.
21. Coalition members also participated in monthly General Coalition meetings. The coalition's Steering Committee was made up of cochairs of the coalition's policy committees. Kirsch, interview, October 2011.
22. Few, Brown, and Tompkins, "Public Participation."
23. Bulkeley et al., "Climate Justice." See also Mendez, "Local Climate Action Plans"; Anguelovski and Carmin, "Something Borrowed"; Few, Brown, and Tompkins, "Public Participation." In the course of this research, I surveyed 41 California climate action plans in cities with census tracts ranked in the 20th percentile under the California Environmental Protection Agency's CalEnviroScreen tool and reviewed the literature on the development of such plans.
24. Quoted in Ella Baker Center and OCAC, *Toolkit.*
25. OCAC member, interview, May 2013.
26. Ella Baker Center and OCAC, *Toolkit*; OCAC member, interview, May 2013.
27. OCAC Steering Committee member, interview, May 2013.
28. Beveridge, interview, June 2013.
29. Quoted in Ella Baker Center and OCAC, *Toolkit.*
30. OCAC Steering Committee member, interview, August 2013.
31. OCAC member, interview, May 2013.
32. Garzón et al., *Community-Based Climate Adaptation Planning.* Community-based participatory research integrates the technical expertise of researchers with the experiential knowledge of non–academically trained community partners who are directly affected by the issue being studied. Minkler and Wallerstein, *Community-Based Participatory Research.*
33. The California Energy Commission and San Francisco Foundation grants were awarded in 2010, a year after the city started its climate action plan process. As a result, the Oakland climate action plan references the coalition's community-based adaptation study as a priority that the city will support in the near future. The climate action plan is updated

every five years, and it is anticipated that the coalition study will be incorporated and expanded in the next update.

34. Garzón, interview, June 2013. See also Garzón et al., *Community-Based Climate Adaptation Planning*.
35. Garzón et al., *Community-Based Climate Adaptation Planning*; OCAC Steering Committee member, interview, August 2013.
36. Martello, "Arctic Indigenous Peoples"; Irwin, *Citizen Science*, 45.
37. OCAC member, interview, May 2013.
38. Martello, "Arctic Indigenous Peoples"; Garzón et al., *Community-Based Climate Adaptation Planning*.
39. Miller, "New Civic Epistemologies," 405.
40. Miller, "New Civic Epistemologies," 406; Iles, "Identifying Environmental Health Risks."
41. Martello and Jasanoff, "Globalization and Environmental Governance," 5; Martello, "Paradox of Virtue?"
42. Martello, "Paradox of Virtue?"
43. Jasanoff, "New Climate for Society," 249; Corburn, "Cities, Climate Change," 414–15.
44. Jasanoff, "New Climate for Society," 236.

Chapter 5. Cap and Trade-Offs

1. Agyeman et al., "Trends and Directions."
2. For purposes of this chapter, I use the name Climate Change Community Benefits Fund. However, under state statute, the fund is referred to as the Greenhouse Gas Reductions Fund. California Health and Safety Code § 39710.
3. I use the term "regulatory culture" to describe the shared norms, values, and beliefs that influence the behavior of staff working within a regulatory agency. According to Sherwin, Davenport, and Scott, "These norms, values and beliefs are influenced by several factors: (1) the beliefs, values, assumptions and behaviors of the founding leaders of the organization; (2) the shared experiences of staff in the performance of the duties; and (3) new beliefs, values, and assumptions brought in by new staff, particularly new leaders." *Regulatory Institutions and Practices*, 77.
4. Kirsch, interview, October 26, 2011.
5. Taruc, interview, August 2017.
6. Abers and Keck, *Practical Authority*, 6–7.
7. Meidinger, "Regulatory Culture," 356–57. See also Knox-Hayes, *Cultures of Markets*, 29–30, 40.
8. Knox-Hayes, *Cultures of Markets*, 15.
9. Abers and Keck, *Practical Authority*, 17–18.
10. Abers and Keck, *Practical Authority*, 17–18.
11. Knox-Hayes, *Cultures of Markets*, 42.

12. EAAC, *Allocating Emissions Allowances*; *Inside Cal/EPA*, "Carbon Revenue Bill."
13. CARB, "AB 32 Scoping Plan Functional Equivalent Document." See also Liévanos, "Certainty, Fairness," for further discussion on issues of fairness in regulatory science.
14. Environmental justice advocate, interview, May 2013; senior Schwarzenegger appointee, interview, August 2013.
15. Carmichael and Prasad, *AB 32 Community Benefit Fund*; environmental justice advocate, interview, May 2013; senior Schwarzenegger appointee, interview, August 2013. See also Kunkel and Kammen, "Carbon Cap and Dividend Policies."
16. Senior capitol staff member, interview, June 2013; senior Schwarzenegger appointee, interview, August 2013. The Association of Irritated Residents filed its lawsuit against the board on June 11, 2009.
17. Senior capitol staff member, interview, June 2013.
18. De León, "Sets Sights on Cleaning-Up."
19. Parino, interview, July 2013.
20. California Health and Safety Code §§ 38505(d), 38560, 38561, 38565.
21. Taruc, interview, August 2017; environmental justice advocate, interview, May 2013.
22. CMTA, "Cap-and-Trade Auction."
23. Environmental justice advocate, interview, May 2013.
24. *Inside Cal/EPA*, "Carbon Revenue Bill."
25. Holmes-Gen, "Calif. Bill."
26. Senior capitol staff member, interview, June 2013.
27. Parino, interview, July 2013.
28. Environmental justice advocate, interview, May 2013; traditional environmental organization member, August 2013. Coalition members include the Center on Race, Poverty, and the Environment; the Asian Pacific Environmental Network; Communities for a Better Environment; the Center for Community Action and Environmental Justice; the Environmental Health Coalition; and People Organizing to Demand Environmental and Economic Rights.
29. Senior capitol staff member, interview, June 2013. De León was appointed chair on June 13, 2008. The Assembly Appropriations Committee reviews all bills with any fiscal impact after passage by a policy committee, with the goal of sound, fiscally responsible policy. Its chairperson has enormous power to bring home special projects for his or her district, as well as having the final say on whether a member's legislative proposal is "too costly" for the state to implement. Senior capitol staff member, interview, June 2013.
30. Senior capitol staff member, interview, June 2013.
31. California Legislative Information, "AB-1405."
32. California Channel, "Assembly Floor Vote on AB 1405."

33. *Inside Cal/EPA*, "California Community Fund."
34. California Channel, "Assembly Floor Vote on AB 1405."
35. *Inside Cal/EPA*, "California Community Fund"; Schwarzenegger, "Governor's Veto Message"; California Legislative Information, "AB-1405."
36. Senior Schwarzenegger appointee, interview, August 2013. In June 2010, the board had recently released a plan that detailed a number of pollution reduction goals for railway operators, but environmental justice activists claimed that the proposals were too weak. According to Penny Newman, executive director of the Center for Community Action and Environmental Justice, "We had a march in front of Mary Nichols' house to bring home to her how serious the situation is. . . . I think people are outraged. You can't stand by and continue to let people be harmed. It's killing people. It's literally killing people." Quoted in Leung, "Residents Protest"; *Inside Cal/EPA*, "ARB Advisers Favor Carbon Revenue."
37. Senior capitol staff member, interview, June 2013.
38. Traditional environmental organization member, August 2013.
39. Newtown, *LAO's Critique*, 8. SB 535 passed in the senate with 23 votes on June 2, 2011. Two members of the Latino Legislative Caucus, Senators Lou Correa and Gloria Negrete McLeod, voted against it. Both represent moderate suburban districts in Southern California. California Legislative Information, unofficial ballot on SB 535.
40. See Liévanos, "Certainty, Fairness," for further discussion of certainty claims made by industry during regulatory proceedings.
41. In a private meeting, the chair of the committee, Assembly Member Felipe Fuentes, informed de León that SB 535 would not be considered for passage. According to several senior capitol sources, the committee held SB 535 under direct instructions of Assembly Speaker John Pérez. The speaker was allegedly upset that de León had withdrawn his support for one of his high-profile bills; in retaliation, Pérez ordered SB 535 held in the Appropriations Committee. Senior capitol staff member, interview, October 2013.
42. Sadd et al., "Playing It Safe," 1443.
43. CEJA, "Comment Letter." See also Sadd et al., "Playing It Safe."
44. CEJA, "Comment Letter." See also Sadd et al., "Playing It Safe."
45. CEJA, "Comment Letter"; environmental justice advocate, interview, May 2013.
46. Corburn, *Street Science*, 176–78; Scott, *Seeing like a State*.
47. Sadd et al., "Playing It Safe," 1445.
48. OEHHA, *Environmental Health Screening Tool*; Alexeeff et al., "Screening Method"; senior Schwarzenegger appointee, interview, August 2013. For more information on the implications of excluding race/ethnicity in the tool, see Liévanos, "Retooling CalEnviroScreen."
49. CEJA, "Comment Letter."
50. Senior capitol staff member, interview, October 2013. See also Liévanos,

"Certainty, Fairness," for further discussion on issues of validity in regulatory science.

51. *Inside Cal/EPA*, "Activists, Industry at Odds."
52. California Health and Safety Code § 39711.
53. Taruc, interview, August 2017.
54. Senior capitol staff member, interview, October 2013.
55. Rodriquez, comments.
56. Senior capitol staff member, interview, October 2013; *Inside Cal/EPA*, "Contentious Bill."
57. According to capitol sources, Assembly Speaker Pérez imposed these conditions as retribution on de León for derailing his legislation the previous year. Rather than rejecting SB 535 in committee, Pérez permanently linked himself to the legacy of California's climate fund by taking away a key provision from de León. This reprisal also denied de León credit as the only legislator responsible for establishing the "first-in-nation" fund. Senior capitol staff member, interview, October 2013; environmental justice advocate, interview, May 2013.
58. California Health and Safety Code §§ 38505(d), 38560, 38561, 38565. As noted in Chapter 3, on October 20, 2011, the board formally adopted a cap-and-trade program. The first allowance auction was scheduled to take place in November 2012.
59. Ella Baker Center for Human Rights, "A Win for California!"
60. Environmental justice advocate, interview, May 2013. While environmental and social justice groups supported Pérez's AB 1532, they did not cosponsor it. Perhaps in a sign of loyalty to Senator de León, they referred to themselves only as the SB 535 Coalition.
61. Environmental justice advocate, interview, May 2013.
62. Environmental justice advocate, interview, May 2013.
63. Environmental justice advocate, interview, May 2013.
64. TransForm and California Housing Partnership Corporation, *Creating and Preserving Affordable Homes.*
65. Environmental justice advocate, interview, May 2013; Abers and Keck, *Practical Authority.*
66. CARB, *Annual Report to the Legislature*; Davis and Smith, *Investments to Benefit Disadvantaged Communities.*
67. Taruc, interview, August 2017.
68. Abers and Keck, *Practical Authority*, 3–4.
69. Barab et al., "Relating Narrative," 61–62.
70. Bullard, "Environmental Justice," 156. See also London, Sze, and Liévanos, "Problems, Promise."
71. Knox-Hayes, *Cultures of Markets.*
72. Jasanoff, "Cosmopolitan Knowledge."

Chapter 6. Climate beyond Borders

1. "Global South" often refers to all nations in Africa, Asia, Latin America, and the Caribbean that the World Bank classifies as low and middle income. Mitlin and Satterthwaite, *Urban Poverty in the Global South*, 13.
2. Young, "Schwarzenegger Opens Climate Summit." Scientists argue that preserving trees is critical because they capture and store carbon. When forests are burned or cut down, however, the carbon contained in these trees is released, and the capacity for sequestering carbon emissions is lost. Wold, Hunter, and Powers, "Land Use and Forestry."
3. Young, "Schwarzenegger Opens Climate Summit."
4. Young, "Schwarzenegger Opens Climate Summit."
5. Schwarzenegger, "Memorandum of Understanding"; EDF, "Tropical Forests Protection Pact."
6. Greiner and Sakdapolrak, "Translocality," 373; McFarlane, "Translocal Assemblages," 561. See also Miller, "Political Empowerment"; Nicholls, Miller, and Beaumont, *Spaces of Contention*.
7. Peth, "What Is Translocality?" See also Pellow, *Resisting Global Toxics*.
8. Protestors were from communities in the region of Amador Hernández and the municipalities of Las Margaritas, Marqués de Comillas, San Juan Cancuc, Oxchuc, and Chenalhó in the state of Chiapas, Mexico. Conant, interview, July 19, 2013.
9. Quoted in Friends of the Earth, *Faces of Change*, 6–7; Conant, interview, July 19, 2013.
10. Conant, interview, July 19, 2013. See also Osborne, "Fixing Carbon."
11. Schlosberg and Carruthers, "Indigenous Struggles," 18.
12. Kelly and Peluso, "Frontiers of Commodification." See also Cabello and Gilbertson, "Colonial Mechanism"; Dressler and McDermott, "Indigenous Peoples and Migrants"; Naughton-Treves and Day, *Lessons about Land Tenure*.
13. Osborne, "Tradeoffs in Carbon Commodification"; Aguilar at CARB, meeting, October 18, 2012; Lueders et al., *California REDD+ Experience*, 15; Hall, *Forests and Climate Change*; Greenpeace, *Outsourcing Hot Air*.
14. Blue Source, the Forestland Group, and Shell, "California Carbon Offset Transaction"; Zuckerman and Koenig, *From Well to Wheel*, 4; Williams et al., "Systems Thinking," 866.
15. Quoted in Halper, "California Fighting Pollution."
16. Environmental justice advocate, interview, August 2017.
17. Lansing et al., review symposium, 2391; Lohmann, "Performative Equations," 170–73.
18. Agarwal and Narain, *Global Warming*. See also Picq, "REDD Trap"; Osborne, "Fixing Carbon."
19. Taruc, interview, August 2017.
20. Taruc, interview, August 2017.

21. Lang, "Protests in Chiapas."
22. Porta and Diani, *Social Movements*, 14. The anthropologist Jeffery Juris also argues that these translocal movements encourage a logic of vertical networking. Evident in many translocal anticorporate organizations, including the World Social Forum and People's Global Action, such networking offers a democratizing principle within social movements. Juris, *Networking Futures*.
23. Granovetter, "Strength of Weak Ties."
24. Keck and Sikkink, *Activists beyond Borders*, 241.
25. Lim, "Cyber-Urban Space Odyssey," 121.
26. Quoted in Conant, "Should Chiapas Farmers Suffer." See also Keck and Sikkink, *Activists beyond Borders*.
27. McFarlane, "Translocal Assemblages," 566; Conant, interview, July 19, 2013.
28. No REDD Tour, "Indigenous Peoples."
29. Lim, "Cyber-Urban Space Odyssey," 121; Governors' Climate and Forests Task Force, "About."
30. Environmental justice advocate, interview, August 2017.
31. Environmental justice advocate, interview, August 2017.
32. CARB, meeting, October 18, 2012.
33. Aguilar at CARB, meeting, October 18, 2012.
34. Aguilar at CARB, meeting, October 18, 2012.
35. Ninawa at CARB, meeting, October 18, 2012.
36. Parino at CARB, meeting, October 18, 2012.
37. Environmental justice advocate, interview, May 2013.
38. No REDD Tour, "Indigenous Peoples." The national Bioneers Conference highlights the work of scientific and social innovators and helps support and propagate their ideas and models.
39. Ramo, "Comments."
40. Environmental justice advocate, interview, May 2013; senior capitol staff member, interview, August 2013.
41. Sunderlin et al., "REDD+ at a Critical Juncture," 5–8.
42. Brown, *Redeeming REDD*, 162.
43. Morain, "Cool Bid"; O'Brien and Kempf, "CARB Invalidates Offsets"; CARB, "Compliance Offset Protocol." Before the board invalidated some of its offsets, Clean Harbors saw increasing demand for the incineration of ozone-depleting substances, largely because of California's cap-and-trade program. In 2014, it was estimated that Clean Harbors managed 87 percent of the 5.5 million credits related to ozone-depleting substances issued by the board. Lusvardi, "Pollution-Offset Credits." In the same year, Clean Harbors announced plans to invest $100 million to build an additional rotary kiln incinerator, which would nearly double the El Dorado facility's hazardous waste capacity. According to a company spokesperson, this is the "first time in 30 years that a new incinerator has

been built in the United States." *Arkansas Business*, "Clean Harbors to Expand."

44. Halper, "California Fighting Pollution"; Morain, "Cool Bid." The board had initially issued a preliminary notice of its plans to review 1.3 million credits issued between the years 2009 and 2012, when Clean Harbors was deemed potentially noncompliant. The invalidation of these credits potentially could have cost California firms up to $13 million. Lusvardi, "Pollution-Offset Credits."
45. Environmental justice advocate, interview, May 2013.
46. Quoted in Gonzalez, "International REDD."
47. Gonzalez, "Could California Make or Break REDD?"
48. Quoted in Ostrander, "California's Cap and Trade."
49. Bassano, "Boomerang Pattern," 25–26; Keck and Sikkink, *Activists beyond Borders*.
50. The idea of the boomerang pattern is also consistent with the concept of "political opportunity structures." This term refers to the specific features of a political system (e.g., a state, city, or country) that can explain the different successes, strategies, action repertoires, levels of mobilization, and organizational structures of environmental movements in individual countries. Van Der Heijden, "Globalization, Environmental Movements," 28. Peter Eisinger found that the incidence of protest was related to the nature of a city's "political opportunity structure," which he defined as "the degree to which groups are likely to be able to gain access to power and to manipulate the political system." "Conditions of Protest Behavior," 25. For the distinction between two basic political opportunity structures, repression and facilitation, see Tarrow, *Power in Movement*, 92. In general, authoritarian states are viewed to be more inclined to repress social movements, whereas representative ones facilitate them.
51. Gareau, *From Precaution to Profit*, 29–30.
52. Swyngedouw and Heynen, "Urban Political Ecology," 912.
53. Lueders et al., *California REDD+ Experience*. At the time, other traditional environmental groups, such as the Sierra Club and the Planning and Conservation League, supported SB 605. The Natural Resources Defense Council supported the bill after its scope was later restricted to offsets within only the United States.
54. McAfee, "Selling Nature to Save It?" See also Halper, "California Fighting Pollution."
55. Holmes, "Conservation's Friends." See also MacDonald, "Devil."
56. O'Connor, testimony.
57. Environmental justice advocate, interview, August 2017. A review of the board's offset database in 2017 partially supported the environmental justice groups' concerns. Over 76 percent of offset projects were located outside California, and the largest greenhouse gas emitters (such as Shell Oil) were often more likely to use offset credits to meet their AB 32 obli-

gations. CARB, "Compliance Offset Protocol." See also Cushing et al., "Carbon Trading."

58. Four Twenty Seven, *Analysis*.
59. Quoted in Conant, "Outsourcing Global Warming Solutions," 80.
60. Johnson, *California, Acre and Chiapas*.
61. Rosenhall, "Three Fined."
62. Ramo, "Comments."
63. Former senior capitol staff member, interview, October 2014. Rosenhall, "Mercury Lobbying Firm." Lara served several years as director of Núñez's assembly district office in Los Angeles. The former speaker was also an early and strong supporter of Lara's first bid for elective office. Former senior capitol staff member, interview, October 2014.
64. Former senior capitol staff member, interview, October 2014.
65. Barboza, "Yurok Tribe."
66. Quoted in Barboza, "Yurok Tribe."
67. Quoted in New Forests and the Yurok Tribe, "New Forests and Yurok Tribe."
68. Cart, "Carbon Cutters on Edge."
69. CARB, "Offset Project Listing Requirements."
70. Environmental justice advocate, interview, August 2017.
71. Cattelino, "Double Bind." See also Castellanos, Gutiérrez Nájera, and Aldama, *Comparative Indigeneities*; environmental justice advocate, interview, August 2017.
72. In 2012, the Environmental Defense Fund also invited Senator Kevin de León and Senator Lou Correa to visit REDD-type projects in Chiapas, Mexico. In 2009, de León's AB 1404, which would have limited the use of offsets, was passed by the legislature but vetoed by Governor Schwarzenegger. Senator Correa, conversely, has been a strong proponent of offsets and used his Senate Select Committee on California-Mexico Cooperation in 2012 to promote AB 32 linkages with Chiapas. Lopez, "Crossing Borders"; Lueders et al., *California REDD+ Experience*.
73. Short-lived climate pollutants also include tropospheric ozone, methane, and hydrofluorocarbons. Zaelke et al., "Primer on Short-Lived Climate Pollutants." See also Grahame, Klemm, and Schlesinger, "Public Health."
74. Zaelke et al., "Primer on Short-Lived Climate Pollutants"; CARB, *First Update*; US EPA, "Black Carbon Effects."
75. Grahame, Klemm, and Schlesinger, "Public Health"; US EPA, "Black Carbon Effects."
76. California Health and Safety Code § 39730(a)(4). As noted in Chapter 5, this tool was established by Senator Kevin de León's SB 535, which was signed into law in 2013.
77. Tarrow, *Power in Movement*; Bacchetta, El-Tayeb, and Haritaworn, "Queer of Colour Formations," 773.
78. Fischer, "How Much Jet Fuel."

79. Environmental justice advocate, interview, August 2017; Harvey, *New Imperialism*. For further discussion on issues of colonialism, see Bacon, "Settler Colonialism"; Fenelon, "Critique of Glenn."
80. Quoted in Hayden, "Gov. Jerry Brown."
81. Senior capitol staff member, interview, August 2017.
82. Quoted in Kahn, "Golden State Climate Diplomacy."
83. Florez, comments.
84. Friends of the Earth, "Declaration from Acre to California."
85. Global Alliance against REDD, "Forest Peoples"; environmental justice advocate, interview, August 2017.
86. Environmental justice advocate, interview, August 2017.

Chapter 7. California Complexity and Possibility

1. Hulme, "Geographical Work," 7; Morris, "Fear of Dying"; San Joaquin Valley Air Pollution Control District, *Presidential Transition White Paper*, 1.
2. The Clean Air Act, adopted in 1963, regulates emissions and sets penalties for failing to meet national air quality standards. After 15 states led by California's attorney general, Xavier Becerra, sued the US Environmental Protection Agency, arguing that the delay was illegal, the Trump administration reversed its action. Naylor, "White House Reverses"; Silverstein, "New Ozone Pollution Rules."
3. The cap-and-trade reauthorization was signed into law through AB 398. McGreevy, "He Rallied Support"; environmental justice advocate, interview, August 2017; Barboza and Megerian, "Questions Remain."
4. These new rules apply to emitters in federal nonattainment areas. A nonattainment area is a region considered to have air quality worse than the National Ambient Air Quality Standards as defined in the Clean Air Act Amendments of 1970 (Pub. L. No. 91–604, § 109); Kaswan, "Broader Vision for Climate Policy," 119.
5. Kaswan, "Broader Vision for Climate Policy," 119.
6. Karvonen, *Politics of Urban Runoff*, 188.
7. Quoted in Barboza and Megerian, "Questions Remain."
8. London, Sze, and Liévanos, "Problems, Promise," 256; Cohen and Ottinger, "Environmental Justice," 1–2; Cole and Foster, *From the Ground Up*.
9. Harrison, "Coopted Environmental Justice?" See also Pulido, Kohl, and Cotton, "State Regulation and Environmental Justice"; London, Sze, and Liévanos, "Problems, Promise"; Liévanos, "Certainty, Fairness."
10. Harrison, *Pesticide Drift*, 192.
11. In the first four years of operation, California invested more than $2.4 billion from cap-and-trade proceeds into mitigation projects ranging from low-carbon transportation, renewable energy, and urban greening to zero-emissions vehicles. These projects have attracted over $8.2 billion from other funding sources, which represents an average of six dol-

lars leveraged for every dollar invested. Fifty-one percent of the $2.4 billion in implemented projects is also providing benefits to environmental justice communities, including 31 percent going to projects located within these communities. CARB, *Annual Report to the Legislature*.

12. London, Sze, and Liévanos, "Problems, Promise," 289.
13. Hulme, "Geographical Work"; Jasanoff, "Ordering Knowledge, Ordering Society," 38–40.
14. Quoted in Kahn, "What Schwarzenegger Learned."
15. Goldstein, "Weakness of Tight Ties"; Curnutte and Testa, "Consuming Genomes"; Pelling, *Adaptation to Climate Change*, 4. See also Hulme, "Geographical Work."
16. Pellow, *What Is Critical Environmental Justice?*; Holifield, "Framework," 269. See also Lerner, *Sacrifice Zones*; Hooks and Smith, "Treadmill of Destruction." The title of Lerner's book is derived from "National Sacrifice Zones," a term coined by federal government officials to designate areas highly contaminated as a result of mining and processing of uranium into nuclear weapons.
17. Pellow, "Politics by Other Greens," 262–63. See also Pellow, *What Is Critical Environmental Justice?*; Pellow, "Critical Environmental Justice Studies."
18. Guinier and Torres, *Miner's Canary*, 11.
19. Pellow, "Politics by Other Greens," 262; Lerner, *Sacrifice Zones*.
20. Bullard, "Environmental Justice."
21. Cohen and Ottinger, "Environmental Justice"; Tesh and Williams, "Identity Politics." See also Corburn, *Street Science*; DeBacker et al., *Pathways to Resilience*, 53–54.
22. Quoted in Johnson, "Ricardo Lara Grew Up."
23. Briggs, "Use of Indigenous Knowledge," 111; Haraway, "Situated Knowledges."
24. McAdam, McCarthy, and Zald, "Opportunities, Mobilizing Structures."
25. Porta and Diani, *Social Movements*.
26. Ezrahi, *Descent of Icarus*, 69.
27. Van House, "Actor-Network Theory."
28. Abers and Keck, *Practical Authority*, 18.
29. Miller, "New Civic Epistemologies."
30. Hulme, "Geographical Work," 9.
31. Carter, "Environmental Justice 2.0," 11. See also Brulle and Pellow, "Environmental Justice"; Brown et al., "Embodied Health Movements"; Mohai, Pellow, and Roberts, "Environmental Justice."
32. Towers, "Political Geography of Scale," 31–32.
33. CEJA, "Supporting Haiyan Relief."
34. CEJA, "Supporting Haiyan Relief."
35. Grahame, Klemm, and Schlesinger, "Public Health"; Kassel and Annotti, *Dumping Dirty Diesels*.
36. Kassel and Annotti, *Dumping Dirty Diesels*.

Afterword

1. Melendez, "Global Climate Action Summit."
2. Kahn, "Tropical States Move Forward."
3. Kahn, "Tropical States Move Forward."
4. Chemnick, "Calif. Summit a Success."
5. Pastor, *State of Resistance*.

Bibliography

Abers, Rebecca Neaera, and Margaret E. Keck. *Practical Authority: Agency and Institutional Change in Brazilian Water Politics*. Oxford: Oxford University Press, 2013.

Adler, Ben. "In New Calif. Legislature, Moderate Dems Hold Leverage." *Capitol Public Radio*, January 22, 2013. http://archive2.capradio.org/articles/2013/01/22/in-new-calif-legislature,-moderate-dems-hold-leverage.

Agarwal, Anil, and Sunita Narain. *Global Warming in an Unequal World: A Case of Environmental Colonialism*. New Delhi: Centre for Science and Environment, 1991.

Agyeman, Julian, Robert D. Bullard, and Bob Evans. *Just Sustainabilities: Development in an Unequal World*. Cambridge, MA: MIT Press, 2003.

Agyeman, Julian, David Schlosberg, Luke Craven, and Caitlin Matthews. "Trends and Directions in Environmental Justice: From Inequity to Everyday Life, Community, and Just Sustainabilities." *Annual Review of Environment and Resources* 41 (2016): 321–40.

Alber, Gotelind, and Kristine Kern. "Governing Climate Change in Cities: Modes of Urban Climate Governance in Multi-Level Systems." Paper presented at the OECD Conference on Competitive Cities and Climate Change, Milan, Italy, October 2008.

Alexeeff, George V., John B. Faust, Laura Meehan August, Carmen Milanes, Karen Randles, Lauren Zeise, and Joan Denton. "A Screening Method for Assessing Cumulative Impacts." *International Journal of Environmental Research and Public Health* 9, no. 2 (February 2012): 648–59.

Allenby, Braden R., and Daniel Sarewitz. *The Techno-human Condition*. Cambridge, MA: MIT Press, 2013.

Amin, Ash, and Nigel Thrift. *Cities: Reimagining the Urban*. Cambridge: Polity, 2002.

Andrew, Jane, Mary A. Kaidonis, and Brian Andrew. "Carbon Tax: Challenging Neoliberal Solutions to Climate Change." *Critical Perspectives on Accounting* 21, no. 7 (October 2010): 611–18.

Angotti, Tom, and Julie Sze. "Environmental Justice Praxis: Implications for Interdisciplinary Urban Health." In *Urban Health and Society: Interdisciplinary Approaches to Research and Practice*, edited by Nicholas Freudenberg, Susan Klitzman, and Susan Saegert, 19–41. San Francisco: Jossey-Bass, 2009.

Anguelovski, Isabelle, and JoAnn Carmin. "Something Borrowed, Everything New: Innovation and Institutionalization in Urban Climate Governance." *Current Opinion in Environmental Sustainability* 3, no. 3 (May 2011): 169–75.

Arkansas Business. "Clean Harbors to Expand in El Dorado, Add 120 Jobs." September 23, 2014. http://www.arkansasbusiness.com/article/100960/clean-harbors-to-expand-in-el-dorado-add-120-jobs.

Avila, Esther. "Bad Air Quality Not Going Away." *Portville Recorder*, January 9, 2014. http://www.recorderonline.com/news/bad-air-quality-not-going-away/article_7e3363c2-78cf-11e3-a791-001a4bcf6878.html.

Bacchetta, Paola, Fatima El-Tayeb, and Jin Haritaworn. "Queer of Colour Formations and Translocal Spaces in Europe." *Environment and Planning D: Society and Space* 33, no. 5 (October 2015): 769–78.

Bacon, J. M. "Settler Colonialism as Eco-social Structure and the Production of Colonial Ecological Violence." *Environmental Sociology* 5, no. 1 (2019): 59–69.

Bailey, Diane, Kim Knowlton, Miriam Rotkin-Ellman, Harris Epstein, Andrew Hoerner, and Gina Solomon. "Boosting the Benefits: Improving Air Quality and Health by Reducing Global Warming Pollution in California." NRDC Issue Paper, June 2008. https://assets.nrdc.org/sites/default/files/boosting.pdf?_ga=2.205059541.459466768.1529242399-814640155.1529242399.

Baptist, Edward. "What the *Economist* Doesn't Get about Slavery—and My Book." *Politico Magazine*, September 7, 2014. https://www.politico.com/magazine/story/2014/09/economist-review-slavery-110687.html.

Barab, Sasha A., Troy D. Sadler, Conan Heiselt, Daniel Hickey, and Steven Zuiker. "Relating Narrative, Inquiry, and Inscriptions: Supporting Consequential Play." *Journal of Science Education and Technology* 16, no. 1 (February 2007): 59–82.

Barboza, Tony. "Heat, Drought Worsen Smog in California, Stalling Decades of Progress." *Los Angeles Times*, November 10, 2014. http://www.latimes.com/science/la-me-air-pollution-20141110-story.html.

———. "Yurok Tribe Hopes California's Cap-and-Trade Can Save a Way of Life." *Los Angeles Times*, December 16, 2014. http://www.latimes.com/science/la-me-carbon-forest-20141216-story.html.

Barboza, Tony, and Chris Megerian. "Questions Remain as Gov. Brown Signs Legislation to Address Neighborhood-Level Air Pollution." *Los Angeles Times*, July 26, 2017. http://www.latimes.com/local/lanow/la-me-ln-air-pollution-law-20170725-story.html.

Barnes, Jessica, and Michael R. Dove, eds. *Climate Cultures: Anthropological Perspectives on Climate Change*. New Haven, CT: Yale University Press, 2015.

Barrett, Sam. "The Necessity of a Multiscalar Analysis of Climate Justice." *Progress in Human Geography* 37, no. 2 (April 2013): 215–33.

Barugh, Hannah, and Dan Glass. "Environment, Climate Change and Popular Education." *Concept: The Journal of Contemporary Community Education Practice Theory* 1, no. 3 (Winter 2010): 1–7.

Bassano, David. "The Boomerang Pattern: Verification and Modification." *Peace & Change* 39, no. 1 (January 2014): 23–48.

Bauer, Greta R. "Incorporating Intersectionality Theory into Population Health Research Methodology: Challenges and the Potential to Advance Health Equity." *Social Science & Medicine* 110 (June 2014): 10–17.

Beck, Ulrich. *World at Risk*. Cambridge: Polity, 2007.

Bentham, Jeremy. *An Introduction to the Principles of Morals and Legislation*. 1780. Reprint of the 1907 edition. Mineola, NY: Dover, 2007.

Bernstein, Sharon. "California Democrats Likely to Stay Moderate Course despite Majority." Reuters, September 13, 2013. https://www.reuters.com/article/us-usa-california-politics/california-democrats-likely-to-stay-moderate-course-despite-majority-idUSBRE98C0ZI20130913.

Berton, Justin, and Kevin Fagan. "Refinery Warning Worked—Mostly." *SFGate*, August 7, 2012. http://www.sfgate.com/bayarea/article/Refinery-warning-worked-mostly-3770326.php.

Beveridge, Brian. Interview by the author, June 2013.

Blue Source, the Forestland Group, and Shell. "Blue Source, the Forestland Group and Shell Complete 500,000 Tonne California Carbon Offset Transaction." Press release, March 15, 2013. http://www.forestlandgroup.com/2013/03/15/complete-500000-tonne-california-carbon-offset-transaction/.

Boyce, James K. Memorandum to Economic and Allocation Advisory Committee regarding investment in disadvantaged communities. October 5, 2009. http://www.climatechange.ca.gov/eaac/documents/member_materials/Boyce_memo_on_investment_in_disadvantaged_communities.pdf.

Boyce, James K., and Manuel Pastor. "Clearing the Air: Incorporating Air Quality and Environmental Justice into Climate Policy." *Climatic Change* 120, no. 4 (October 2013): 801–14.

Brace, Catherine, and Hilary Geoghegan. "Human Geographies of Climate Change: Landscape, Temporality, and Lay Knowledges." *Progress in Human Geography* 35, no. 3 (June 2011): 284–302.

Brah, Avtar, and Ann Phoenix. "Ain't I a Woman? Revisiting Intersectionality." *Journal of International Women's Studies* 5, no. 3 (May 2004): 75–86.

Braveman, Paula A. "Monitoring Equity in Health and Healthcare: A Conceptual Framework." *Journal of Health, Population, and Nutrition* 21, no. 3 (September 2003): 181–92.

Braveman, Paula A., Mercedes Dekker, Susan Egerter, Tabashir Sadegh-Nobari, and Craig Pollack. *Housing and Health*. Exploring the Social Determinants of Health 7. Robert Wood Johnson Foundation, May 2011. https://www.rwjf.org/content/dam/farm/reports/issue_briefs/2011/rwjf70451.

Braveman, Paula A., and Susan Egerter. *Overcoming Obstacles to Health: Report from the Robert Wood Johnson Foundation to the Commission to Build a Healthier Amer-*

ica. Princeton, NJ: Robert Wood Johnson Foundation, February 2008. https://www.rwjf.org/content/dam/farm/reports/reports/2008/rwjf22441.

Brenner, Neil. *New State Spaces: Urban Governance and the Rescaling of Statehood.* Oxford: Oxford University Press, 2004.

———. "State Territorial Restructuring and the Production of Spatial Scale: Urban and Regional Planning in the Federal Republic of Germany, 1960–1990." *Political Geography* 16, no. 4 (May 1997): 273–306.

Briggs, John. "The Use of Indigenous Knowledge in Development: Problems and Challenges." *Progress in Development Studies* 5, no. 2 (April 2005): 99–114.

Broad, Garrett M. *More than Just Food: Food Justice and Community Change*. Oakland: University of California Press, 2016.

Brown, Michael I. *Redeeming REDD: Policies, Incentives and Social Feasibility for Avoided Deforestation*. London: Routledge, 2013.

Brown, Phil. *Toxic Exposures: Contested Illnesses and the Environmental Health Movement*. New York: Columbia University Press, 2007.

Brown, Phil, Stephen Zavestoski, Sabrina McCormick, Brian Mayer, Rachel Morello-Frosch, and Rebecca Gasior Altman. "Embodied Health Movements: New Approaches to Social Movements in Health." *Sociology of Health & Illness* 26, no. 1 (January 2004): 50–80.

Brulle, Robert J., and David N. Pellow. "Environmental Justice: Human Health and Environmental Inequalities." *Annual Review of Public Health* 27 (2006): 103–24.

Bryant, Bunyan, and Paul Mohai, eds. *Race and the Incidence of Environmental Hazards: A Time for Discourse*. Boulder, CO: Westview, 1992.

Bukowski, John. "Review of the CalEPA Report Entitled Cumulative Impacts: Building a Scientific Foundation, Public Review Draft, August 19, 2010." September 17, 2010. https://oehha.ca.gov/media/downloads/calenviroscreen/comments/buscocoms.pdf.

Bulkeley, Harriet, JoAnn Carmin, Vanesa Castán Broto, Gareth A. S. Edwards, and Sara Fuller. "Climate Justice and Global Cities: Mapping the Emerging Discourses." *Global Environmental Change* 23, no. 5 (October 2013): 914–25.

Bulkeley, Harriet, Vanesa Castán Broto, and Gareth A. S. Edwards. *An Urban Politics of Climate Change: Experimentation and the Governing of Socio-technical Transitions*. London: Routledge, 2015.

Bullard, Robert D. *Dumping in Dixie: Race, Class, and Environmental Quality*. 3rd ed. London: Routledge, 2000.

———. "Environmental Justice in the 21st Century: Race Still Matters." *Phylon* 49, no. 3–4 (Autumn–Winter 2001): 151–71.

Burayidi, Michael A., ed. *Urban Planning in a Multicultural Society*. Westport, CT: Praeger, 2000.

Burrell, Jenna. "The Field Site as a Network: A Strategy for Locating Ethnographic Research." *Field Methods* 21, no. 2 (May 2009): 181–99.

Burtraw, Dallas, and Michael A. Toman. "Estimating the Ancillary Benefits of Greenhouse Gas Mitigation Policies in the US." In *Ancillary Benefits and Costs of Greenhouse Gas Mitigation*, 481–513. Paris: OECD, 2000.

C40 Cities. "Cities to Set the Pace and Scope of Progress until Any COP21 Deal Kicks In; C40 and Arup Identify Priority Climate Actions for Cities over the Next Five Years." Press release, December 8, 2015. http://www.c40.org/press_releases/press-release-cities-to-set-the-pace-and-scope-of-progress-until-any-cop21-deal-kicks-in-c40-and-arup-identify-priority-climate-actions-for-cities-over-the-next-five-years.

Cabello, Joanna, and Tamra Gilbertson. "A Colonial Mechanism to Enclose Lands: A Critical Review of Two REDD+-Focused Special Issues." *Ephemera: Theory & Politics in Organization* 12, no. 1/2 (May 2012): 162–80.

Çağlar, Ayse, and Nina Glick Schiller. "A Multiscalar Perspective on Cities and Migration: A Comment on the Symposium." *Sociologica*, no. 2 (May–August 2015): 1–9.

Cal. Chamber. "Environmental Justice: Open Dialogue Can Yield Cost-Effective, Sound Policies." CalChamber Advocacy, January 2017. https://advocacy.calchamber.com/wp-content/uploads/policy/issue-reports/Environmental-Justice-2017.pdf.

———. "Job Killers." CalChamber Advocacy. 2006. http://advocacy.calchamber.com/policy/bill-tracking/job-killers/.

California Channel. "Assembly Floor Vote on AB 32." Video on Demand: California Legislative Hearings. August 30, 2006. http://www.calchannel.com/video-on-demand/.

———. "Assembly Floor Vote on AB 1405." Video on Demand: California Legislative Hearings. August 30, 2010. http://www.calchannel.com/video-on-demand/.

California Legislative Information. "AB-1405 California Global Warming Solutions Act of 2006: California Climate Change Community Benefits Fund." Bill History. Updated September 30, 2010. https://leginfo.legislature.ca.gov/faces/billHistoryClient.xhtml?bill_id=200920100AB1405.

———. Unofficial ballot on AB 32, August 31, 2006. http://leginfo.ca.gov/pub/05-06/bill/asm/ab_0001-0050/ab_32_vote_20060831_0405PM_asm_floor.html.

———. Unofficial ballot on SB 535, June 2, 2011. http://www.leginfo.ca.gov/pub/11-12/bill/sen/sb_0501-0550/sb_535_vote_20110602_0147PM_sen_floor.html.

Cama, Timothy. "Report Faults Chevron Safety, Regulators, in 2012 Refinery Fire." *Hill*, January 23, 2015. http://thehill.com/policy/energy-environment/230536-report-faults-chevron-safety-regulators-in-2012-refinery-fire.

Capitol Weekly. "Former Air Board Nominee Lands New Job with EPA." July 18, 2007. http://capitolweekly.net/former-air-board-nominee-lands-new-job-with-epa/.

CARB. "Adaptive Management." 2011. https://www.arb.ca.gov/cc/capandtrade/adaptivemanagement/adaptivemanagement.htm.

———. *Annual Report to the Legislature on California Climate Investments Using Cap-and-Trade Auction Proceeds*. Sacramento, CA: CARB, March 2018. https://www.arb.ca.gov/cc/capandtrade/auctionproceeds/2018_cci_annual_report.pdf.

———. "Appendix H: Public Health Analysis." In *Climate Change Scoping Plan Appendices, Vol. II: Analysis and Documentation*. December 2008. https://www.arb.ca.gov/cc/scopingplan/document/appendices_volume2.pdf.

———. *Climate Change Scoping Plan: A Framework for Change*. December 2008. https://arb.ca.gov/cc/scopingplan/document/adopted_scoping_plan.pdf.

———. "Collaboration on Climate Change: International Agreements." California Climate Change, 2017. http://climatechange.ca.gov/climate_action_team/partnerships.html.

———. "Compliance Offset Protocol Ozone Depleting Substances (ODS) Projects." October 20, 2011. https://www.arb.ca.gov/cc/capandtrade/protocols/ods/ods.htm.

———. "Final Supplement to the AB 32 Scoping Plan Functional Equivalent Document." August 19, 2011. https://www.arb.ca.gov/cc/scopingplan/document/final_supplement_to_sp_fed.pdf.

———. *First Update to the Climate Change Scoping Plan: Building on the Framework*. Sacramento, CA: CARB, May 2014. https://www.arb.ca.gov/cc/scopingplan/2013_update/first_update_climate_change_scoping_plan.pdf.

———. Meeting, Sacramento, California, October 18, 2012. CAL-SPAN. http://cal-span.org/unipage/?site=cal-span&owner=CARB&date=2012-10-18.

———. "Offset Project Listing Requirements for Native American Tribes." Last reviewed October 15, 2015. https://www.arb.ca.gov/cc/capandtrade/offsets/offset-tribes.htm.

Carbon Leadership Forum. "Embodied Carbon Network." 2017. http://carbonleadershipforum.org/embodied-carbon-network/.

Carmichael, Tim, and Shankar Prasad. "AB 32 Community Benefit Fund to Reduce Cumulative and Disproportionate Impacts." Concept paper. Coalition for Clean Air, March 2008.

Carmin, JoAnn, and Julian Agyeman, eds. *Environmental Inequalities beyond Borders: Local Perspectives on Global Injustices*. Cambridge, MA: MIT Press, 2011.

Carmin, JoAnn, and David Dodman. "Engaging Science and Managing Scientific Uncertainty in Urban Climate Adaptation Planning." In *Successful Adaptation to Climate Change: Linking Science and Policy in a Rapidly Changing World*, edited by Susanne C. Moser and Maxwell T. Boykoff, 220–34. Abingdon, UK: Routledge, 2013.

Cart, Julie. "California Becomes First State to Adopt Cap-and-Trade Program." *Los Angeles Times*, October 21, 2011. http://articles.latimes.com/2011/oct/21/local/la-me-cap-trade-20111021.

———. "Carbon Cutters on Edge: Hoping California's Cap-and-Trade Program Survives." *CALmatters*, December 29, 2016. https://calmatters.org/articles/carbon-cutters-on-edge-hoping-californias-cap-and-trade-program-survives/.

Carter, Eric D. "Environmental Justice 2.0: New Latino Environmentalism in Los Angeles." *Local Environment* 21, no. 1 (January 2016): 3–23.

Cartwright, Anton, Susan Parnell, Gregg Oelofse, and Sarah Ward, eds. *Climate Change at the City Scale: Impacts, Mitigation and Adaptation in Cape Town*. Abingdon, UK: Routledge, 2012.

Castellanos, M. Bianet, Lourdes Gutiérrez Nájera, and Arturo J. Aldama, eds.

Comparative Indigeneities of the Américas: Toward a Hemispheric Approach. Tucson: University of Arizona Press, 2012.

Castro, Tony. "Latinos and the Environmental Movement: Change Is Afoot." *NewsTaco*, September 19, 2013. https://newstaco.com/2013/09/19/latinos-and-the-environmental-movement-change-is-afoot/.

Cattelino, Jessica. "The Double Bind of American Indian Need-Based Sovereignty." *Cultural Anthropology* 25, no. 2 (May 2010): 235–62.

CBE. "Richmond: CBE Advocates for a Just Transition from Fossil Fuels to Building a New Healthier and Thriving Economy." 2016. http://www.cbecal.org/organizing/northern-california/richmond/.

CEJA. "Comment Letter to the Office of Environmental Health Hazard and Assessment Regarding the Draft CalEnviroScreen Tool." October 16, 2012.

———. *Environmental Justice Scorecard 2014*. Oakland, CA: CEJA, 2014. http://caleja.org/wp-content/uploads/2015/01/CEJA_scorecard2014.pdf.

———. "Supporting Haiyan Relief Is Part of Our Fight for Climate Justice." 2013. http://caleja.org/2014/01/1144/.

Chemnick, Jean. "Calif. Summit a Success, but Big Tests Still to Come." *Climatewire*, September 17, 2018. https://www.eenews.net/climatewire/stories/1060097141.

Circular Ecology. "Embodied Energy and Carbon." 2017. http://www.circularecology.com/embodied-energy-and-carbon-footprint-database.html.

City of Oakland. "Oakland's Energy and Climate Action Plan." 2011. http://www2.oaklandnet.com/Government/o/PWA/s/SO/OAK025294.

Clark, Judy, and Jonathan Murdoch. "Local Knowledge and the Precarious Extension of Scientific Networks: A Reflection on Three Case Studies." *Sociologia Ruralis* 37, no. 1 (April 1997): 38–60.

Clark, Lara P., Dylan B. Millet, and Julian D. Marshall. "Changes in Transportation-Related Air Pollution Exposures by Race-Ethnicity and Socioeconomic Status: Outdoor Nitrogen Dioxide in the United States in 2000 and 2010." *Environmental Health Perspectives* 125, no. 9 (September 2017): 1–10.

Clarke, Averil Y., and Leslie McCall. "Intersectionality and Social Explanation in Social Science Research." *Du Bois Review: Social Science Research on Race* 10, no. 2 (Fall 2013): 349–63.

Cleaver, Frances. "Paradoxes of Participation: Questioning Participatory Approaches to Development." *Journal of International Development* 11, no. 4 (June 1999): 597–612.

CMTA. "Cap-and-Trade Auction Revenue Bill Moves to Senate Floor." August 27, 2010. http://www.cmta.net/page/legupdate-article.php?legupdate_id=1765.

Cohen, Benjamin, and Gwen Ottinger. "Introduction: Environmental Justice and the Transformation of Science and Engineering." In *Technoscience and Environmental Justice: Expert Cultures in a Grassroots Movement*, edited by Gwen Ottinger and Benjamin Cohen, 1–18. Cambridge, MA: MIT Press, 2011.

Cole, Luke W., and Caroline Farrell. "Structural Racism, Structural Pollution and the Need for a New Paradigm." *Washington University Journal of Law and Policy* 20 (2006): 265–82.

Cole, Luke W., and Sheila R. Foster. *From the Ground Up: Environmental Racism and the Rise of the Environmental Justice Movement*. New York: New York University Press, 2001.

Conant, Jeff. Interview by the author, July 19, 2013.

———. "Outsourcing Global Warming Solutions." *Race, Poverty & the Environment* 18, no. 1 (2011): 77–82.

———. *A Poetics of Resistance: The Revolutionary Public Relations of the Zapatista Insurgency*. Oakland, CA: AK Press, 2010.

———. "Should Chiapas Farmers Suffer for California's Carbon?" *YES! Magazine*, November 13, 2012. http://www.yesmagazine.org/issues/what-would-nature-do/should-chiapas-farmers-pay-the-price-of-californias-carbon.

Corburn, Jason. "Cities, Climate Change and Urban Heat Island Mitigation: Localising Global Environmental Science." *Urban Studies* 46, no. 2 (February 2009): 413–27.

———. "Community Knowledge in Environmental Health Science: Co-producing Policy Expertise." *Environmental Science & Policy* 10, no. 2 (April 2007): 150–61.

———. *Street Science: Community Knowledge and Environmental Health Justice*. Cambridge, MA: MIT Press, 2005.

———. *Toward the Healthy City: People, Places, and the Politics of Urban Planning*. Cambridge, MA: MIT Press, 2009.

Crenshaw, Kimberlé. "Demarginalizing the Intersection of Race and Sex: A Black Feminist Critique of Antidiscrimination Doctrine, Feminist Theory and Antiracist Politics." *University of Chicago Legal Forum* 1989 (1989): 139–67.

Cuff, Daniel. "Record-Tying Streak of Spare the Air Alerts Ends on Tuesday in Bay Area." *Mercury News*, January 12, 2015. https://www.mercurynews.com/2015/01/12/record-tying-streak-of-spare-the-air-alerts-ends-on-tuesday-in-bay-area/.

Cummins, Steven, Sarah Curtis, Ana V. Diez-Roux, and Sally Macintyre. "Understanding and Representing 'Place' in Health Research: A Relational Approach." *Social Science & Medicine* 65, no. 9 (November 2007): 1825–38.

Curnutte, Margaret, and Giuseppe Testa. "Consuming Genomes: Scientific and Social Innovation in Direct-to-Consumer Genetic Testing." *New Genetics and Society* 31, no. 2 (June 2012): 159–81.

Curtis, S. E., and K. J. Oven. "Geographies of Health and Climate Change." *Progress in Human Geography* 36, no. 5 (October 2012): 654–66.

Cushing, Lara, Dan Blaustein-Rejto, Madeline Wander, Manuel Pastor, James Sadd, Allen Zhu, and Rachel Morello-Frosch. "Carbon Trading, Co-pollutants, and Environmental Equity: Evidence from California's Cap-and-Trade Program (2011–2015)." *PLoS Medicine* 15, no. 7 (July 2018): 1–20.

Dalrymple, Jim, II. "California Was Basically Treated like Its Own Country at the Big Climate Talks in Paris." *BuzzFeed News*, December 11, 2015. https://www

.buzzfeednews.com/jimdalrympleii/california-was-basically-treated-like-its-own-country-at-the.

Davidoff, Paul. "Advocacy and Pluralism in Planning." *Journal of the American Planning Association* 31, no. 4 (1965): 331–38.

Davis, Monique Spears, and Bailey Smith. *Investments to Benefit Disadvantaged Communities, Senate Bill 535 (De León, Chapter 830, Statutes of 2012): Cap-and-Trade Auction Proceeds; Interim Guidance to Agencies Administering Greenhouse Gas Reduction Fund Monies*. CARB, November 3, 2014. https://www.arb.ca.gov/cc/capandtrade/auctionproceeds/final535-interim-guidance-11-3-2014.pdf.

Dayaneni, Gopal. "Carbon Fundamentalism vs. Climate Justice." *Race, Poverty & the Environment* 16, no. 2 (Fall 2009): 7–11.

DeBacker, Lois, Shamar A. Bibbins, Marian Urquilla, Tàj James, Jesse Clarke, Laurie Mazur, Heather Boyer, and Rachel Burrows, eds. *Pathways to Resilience: Transforming Cities in a Changing Climate*. Oakland, CA: Movement Strategy Center, January 2015. https://kresge.org/sites/default/files/Pathways-to-resilience-2015.pdf.

De la Cruz, Alegria. "'Cap and Trade Will Fail': An Interview with Alegria de La Cruz of the Center on Race, Poverty and the Environment." By Jeff Conant. *The Watering Hole* (blog), November 17, 2011. https://jeffconant.wordpress.com/2011/11/17/cap-and-trade-will-fail-an-interview-with-alegria-de-la-cruz-of-the-center-on-race-poverty-and-the-environment/.

De León, Kevin. "Senator de Leon Sets Sights on Cleaning-Up California's Most Polluted Neighborhoods." Press release, February 23, 2011. https://ellabakercenter.org/in-the-news/press-release-news-story/senator-de-león-sets-sights-on-cleaning-up-california's-most.

Des Chene, Mary. "Locating the Past." In *Anthropological Locations: Boundaries and Grounds of a Field Science*, edited by Akhil Gupta and James Ferguson, 66–85. Berkeley: University of California Press, 1997.

Dhamoon, Rita Kaur. "Considerations on Mainstreaming Intersectionality." *Political Research Quarterly* 64, no. 1 (March 2011): 230–43.

Diffenbaugh, Noah S., Daniel L. Swain, and Danielle Touma. "Anthropogenic Warming Has Increased Drought Risk in California." *Proceedings of the National Academies of Sciences* 112, no. 13 (March 31, 2015): 3931–36.

Doerr, John. "Why Smart Businesses Stand behind Global Warming Law." *Environmental Defense Solutions* 37, no. 5 (November–December 2006): 5. https://www.edf.org/sites/default/files/5606_Solutions1106.pdf.

Douglas, Ellen M., Paul H. Kirshen, Michael Paolisso, Chris Watson, Jack Wiggin, Ashley Enrici, and Matthias Ruth. "Coastal Flooding, Climate Change and Environmental Justice: Identifying Obstacles and Incentives for Adaptation in Two Metropolitan Boston Massachusetts Communities." *Mitigation and Adaptation Strategies for Global Change* 17, no. 5 (June 2012): 537–62.

Dressler, Wolfram H., and Melanie Hughes McDermott. "Indigenous Peoples and Migrants: Social Categories, Rights, and Policies for Protected Areas in the

Philippine Uplands." *Journal of Sustainable Forestry* 29, no. 2–4 (May 2010): 328–61.

Drury, Richard Toshiyuki, Michael E. Belliveau, J. Scott Kuhn, and Shipra Bansal. "Pollution Trading and Environmental Injustice: Los Angeles' Failed Experiment in Air Quality Policy." *Duke Environmental Law & Policy Forum* 9 (Spring 1999): 231–90.

EAAC. *Allocating Emissions Allowances under a California Cap-and-Trade Program: Recommendations to the California Air Resources Board and California Environmental Protection Agency*. Final report, March 2010. http://www.climatechange.ca.gov/eaac/documents/eaac_reports/2010-03-22_EAAC_Allocation_Report_Final.pdf.

EDF. "California's Clean Cars Law: Renewed Hope for Protective Tailpipe Standards to Cut Global Warming Pollution." 2012.

———. "First in Nation, California Caps Global Warming Pollution." *Environmental Defense Solutions* 37, no. 5 (November–December 2006): 1–2. https://www.edf.org/sites/default/files/5606_Solutions1106.pdf.

———. "'Getting to Yes' on Climate Action." *Environmental Defense Solutions* 37, no. 5 (November–December 2006): 4–5. https://www.edf.org/sites/default/files/5606_Solutions1106.pdf.

———. "How Cap and Trade Works." 2017. https://www.edf.org/climate/how-cap-and-trade-works.

———. "Tropical Forests Protection Pact Hailed for Protecting Climate." Press release, November 16, 2010. https://www.edf.org/news/tropical-forests-protection-pact-hailed-protecting-climate.

Egelko, Bob. "Calif. Cap-and-Trade Plan Suffers Legal Setback." *SFGate*, March 21, 2011. https://www.sfgate.com/news/article/Calif-cap-and-trade-plan-suffers-legal-setback-2387880.php.

Eisinger, Peter K. "The Conditions of Protest Behavior in American Cities." *American Political Science Review* 67, no. 1 (March 1973): 11–28.

EJAC. "Recommendations and Comments of the Environmental Justice Advisory Committee on the Implementation of the Global Warming Solutions Act of 2006 (AB 32) on the Draft Scoping Plan." October 2008. https://www.arb.ca.gov/cc/ejac/ejac_comments_final.pdf.

———. "Recommendations and Comments of the Environmental Justice Advisory Committee on the Implementation of the Global Warming Solutions Act of 2006 (AB 32) on the Proposed Scoping Plan." December 2008. https://www.arb.ca.gov/cc/ejac/proposedplan-ejaccommentsfinaldec10.pdf.

EJ Matters. "The California Environmental Justice Movement's Declaration on Use of Carbon Trading Schemes to Address Climate Change." 2008. http://www.ejmatters.org/declaration.html (website discontinued).

———. "Environmental Justice Advocates File Suit to Force California to Follow the Law in Implementing the AB 32 Global Warming Solutions Act." Press release, June 10, 2009. http://www.ejmatters.org/docs/Scope_Plan_Litigation_PRESS_RELEASE_6-10-09.doc (website discontinued).

———. "To Address Climate Change." 2008. http://www.ejmatters.org/ (website discontinued).

Ella Baker Center for Human Rights. "A Win for California!" *Ella's Voice* (blog), October 1, 2012. http://ellabakercenter.org/blog/2012/10/a-win-for-california-0.

Ella Baker Center for Human Rights and OCAC. *A Toolkit to Create Climate Action in Your Community*. 2012. http://ellabakercenter.org/sites/default/files/downloads/OCAC-Toolkit.pdf.

Environmental justice advocate. Interview by the author, August 2013.

———. Interview by the author, August 2017.

———. Interview by the author, June 2012.

———. Interview by the author, May 2013.

———. Interview by the author, September 2016.

Esty, Daniel C. "Bottom-Up Climate Fix." *New York Times*, September 21, 2014. https://www.nytimes.com/2014/09/22/opinion/bottom-up-climate-fix.html.

Ezrahi, Yaron. *The Descent of Icarus: Science and the Transformation of Contemporary Democracy*. Cambridge, MA: Harvard University Press, 1990.

Faber, Daniel R., and Deborah McCarthy. *Green of Another Color: Building Effective Partnerships between Foundations and the Environmental Justice Movement*. Northeastern University, Philanthropy and Environmental Justice Research Project, April 10, 2001. http://www.northeastern.edu/nejrc/wp-content/uploads/Another-Color-Final-Report.pdf.

Farrell, Caroline. "A Just Transition: Lessons Learned from the Environmental Justice Movement." *Duke Forum for Law & Social Change* 4 (2012): 45–63.

Faust, John, Laura August, George Alexeeff, Komal Bangia, Rose Cendak, Elyse Cheung-Sutton, Lara Cushing, et al. *California Communities Environmental Health Screening Tool, Version 2.0 (CalEnviroScreen 2.0)*. OEHHA, October 2014. https://oehha.ca.gov/media/downloads/report/ces20finalreportupdateoct2014.pdf.

Fenelon, James V. "Critique of Glenn on Settler Colonialism and Bonilla-Silva on Critical Race Analysis from Indigenous Perspectives." *Indigenous Sociologies* 2, no. 2 (April 2016): 237–42.

Few, Roger, Katrina Brown, and Emma L. Tompkins. "Public Participation and Climate Change Adaptation: Avoiding the Illusion of Inclusion." *Climate Policy* 7, no. 1 (2007): 46–59.

First National People of Color Environmental Leadership Summit. "17 Principles of Environmental Justice." 1991. Posted in Environmental Working Group, October 2, 2007. https://www.ewg.org/enviroblog/2007/10/17-principles-environmental-justice.

Fischer, Frank. *Citizens, Experts, and the Environment: The Politics of Local Knowledge*. Durham, NC: Duke University Press, 2000.

———. *Climate Crisis and the Democratic Prospect: Participatory Governance in Sustainable Communities*. Oxford: Oxford University Press, 2017.

Fischer, Frank, Douglas Torgerson, Anna Durnová, and Michael Orsini, eds. *Handbook of Critical Policy Studies*. Cheltenham, UK: Edward Elgar, 2015.

Fischer, Jared. "How Much Jet Fuel Did Arnold Schwarzenegger Waste as Governor?" *Hopes & Fears*, July 2, 2015. http://www.hopesandfears.com/hopes/now/question/214841-how-much-jet-fuel-did-arnold-schwarzenegger-waste.

Fisher, Dana. *National Governance and the Global Climate Change Regime*. Lanham, MD: Rowman & Littlefield, 2004.

Florez, Dean. Comments during the panel "Diverse Voices in Environmental Leadership: California Cap & Trade Laws, Forests, and Indigenous Rights." Yale School of Forestry, New Haven, CT, September 16, 2016.

Forester, John. *Planning in the Face of Power*. Berkeley: University of California Press, 1988.

Former Latino Caucus member. Interview by the author, August 2013.

———. Interview by the author, October 2017.

Former senior capitol staff member. Interview by the author, October 2014.

Fortun, Kim. "Scaling and Visualizing Multi-sited Ethnography." In *Multi-sited Ethnography: Theory, Praxis and Locality in Contemporary Research*, edited by Mark-Anthony Falzon, 73–85. New York: Routledge, 2009.

Four Twenty Seven. *Analysis of the Impact of SB 605 on the California Carbon Market*. Berkeley, CA: Four Twenty Seven, July 30, 2013. http://427mt.com/wp-content/uploads/2013/07/Market-Impact-of-SB-605.pdf.

Friends of the Earth. "Declaration from Acre to California." YouTube video, June 26, 2017. https://www.youtube.com/watch?v=eo-upIzVSus&t=.

———. *The Faces of Change: Annual Report 2012*. Washington, DC: Friends of the Earth, 2012. https://foe.org/resources/annual-report-2012-the-faces-of-change/.

Gareau, Brian J. *From Precaution to Profit: Contemporary Challenges to Environmental Protection in the Montreal Protocol*. New Haven, CT: Yale University Press, 2013.

Garzón, Catalina. Interview by the author, June 2013.

Garzón, Catalina, Heather Cooley, Matthew Heberger, Eli Moore, Lucy Allen, Eyal Matalon, Anna Doty, and OCAC. *Community-Based Climate Adaptation Planning: Case Study of Oakland, California*. CEC-500-2012-038. California Energy Commission, July 2012. https://www.energy.ca.gov/2012publications/CEC-500-2012-038/CEC-500-2012-038.pdf.

Giddens, Anthony. *The Politics of Climate Change*. Cambridge: Polity, 2009.

Glasberg, Davita Silfen, and Deric Shannon. *Political Sociology: Oppression, Resistance, and the State*. Thousand Oaks, CA: Pine Forge, 2011.

Global Alliance against REDD. "Forest Peoples in Brazil Send a Message to California: Reject Tropical Forest Offsets." 2017. http://no-redd.com/reject-tropical-forest-offsets/.

Goldstein, Bruce Evan. "The Weakness of Tight Ties: Why Scientists Almost Destroyed the Coachella Valley Multispecies Habitat Conservation Plan in Order to Save It." *Environmental Management* 46, no. 2 (August 2010): 268–84.

González, Erualdo Romero, Raul P. Lejano, Guadalupe Vidales, Ross F. Conner, Yuki Kidokoro, Bahram Fazeli, and Robert Cabrales. "Participatory Action

Research for Environmental Health: Encountering Freire in the Urban Barrio." *Journal of Urban Affairs* 29, no. 1 (February 2007): 77–100.

Gonzalez, Gloria. "Could California Make or Break REDD?" Forest Carbon Portal, 2013. http://www.forestcarbonportal.com/news/could-california-make-or-break-redd.

———. "International REDD Faces Uphill Battle in California in 2014." *Forest Trends* (blog), October 8, 2013. https://www.forest-trends.org/ecosystem_marketplace/international-redd-faces-uphill-battle-in-california-in-2014/.

Goodale, Gloria. "At Paris Climate Talks, California Has a Message for the World." *Christian Science Monitor*, December 2, 2015. https://www.csmonitor.com/Environment/2015/1202/At-Paris-climate-talks-California-has-a-message-for-the-world.

Governors' Climate and Forests Task Force. "About." 2016. https://gcftf.org/about/.

Graham, Stephen, and Patsy Healey. "Relational Concepts of Space and Place: Issues for Planning Theory and Practice." *European Planning Studies* 7, no. 5 (October 1999): 623–46.

Grahame, Thomas J., Rebecca Klemm, and Richard B. Schlesinger. "Public Health and Components of Particulate Matter: The Changing Assessment of Black Carbon." *Journal of the Air & Waste Management Association* 64, no. 6 (June 2014): 620–60.

Granovetter, Mark S. "The Strength of Weak Ties." *American Journal of Sociology* 78, no. 6 (May 1973): 1360–80.

Greenpeace. *Outsourcing Hot Air: The Push for Sub-national REDD Offsets in California's Carbon Market from Mexico and Beyond*. Amsterdam: Greenpeace International, September 2012. http://www.greenpeace.org/international/Global/international/publications/forests/2012/REDD/OutsourcingHotAir.pdf.

Greiner, Clemens, and Patrick Sakdapolrak. "Translocality: Concepts, Applications and Emerging Research Perspectives." *Geography Compass* 7, no. 5 (May 2013): 373–84.

Grimes, Katy. "Anti-democracy Bill Guts California Open-Government Laws." *CalWatchdog* (blog), July 18, 2012. https://calwatchdog.com/2012/07/18/anti-democracy-bill-guts-california-open-government-laws/.

Guinier, Lani, and Gerald Torres. *The Miner's Canary: Enlisting Race, Resisting Power, Transforming Democracy*. Cambridge, MA: Harvard University Press, 2003.

Gupta, Akhil, and James Ferguson. "Beyond 'Culture': Space, Identity, and the Politics of Difference." In *Culture, Power, Place: Explorations in Critical Anthropology*, edited by Akhil Gupta and James Ferguson, 33–51. Durham, NC: Duke University Press, 1997.

Habibi, Don A. *John Stuart Mill and the Ethic of Human Growth*. Dordrecht, the Netherlands: Springer, 2001.

Hall, Anthony. *Forests and Climate Change: The Social Dimensions of REDD in Latin America*. Cheltenham, UK: Edward Elgar, 2012.

Hallegatte, Stéphane, Franck Lecocq, and Christian de Perthuis. *Designing Climate*

Change Adaptation Policies: An Economic Framework. Policy Research Working Paper 5568. Washington, DC: World Bank, February 2011. http://documents.worldbank.org/curated/en/596081468155974532/Designing-climate-change-adaptation-policies-an-economic-framework.

Halper, Evan. "Activists Say California Fighting Pollution Globally but Not Locally." *Los Angeles Times*, June 29, 2014. http://www.latimes.com/nation/la-fi-climate-credits-20140630-story.html.

Haraway, Donna. "Situated Knowledges: The Science Question in Feminism and the Privilege of Partial Perspective." *Feminist Studies* 14, no. 3 (Fall 1988): 575–99.

Harmon, Steven. "Schwarzenegger Pressured Air-Quality Board, Democrats Charge." *Mercury News*, July 2, 2007. https://www.mercurynews.com/2007/07/02/schwarzenegger-pressured-air-quality-board-democrats-charge/.

Harrison, Jill Lindsey. "Coopted Environmental Justice? Activists' Roles in Shaping EJ Policy Implementation." *Environmental Sociology* 1, no. 4 (October 2015): 241–55.

———. *Pesticide Drift and the Pursuit of Environmental Justice*. Cambridge, MA: MIT Press, 2011.

Harvey, David. *Justice, Nature, and the Geography of Difference*. Oxford: Blackwell, 1996.

———. *The New Imperialism*. Oxford: Oxford University Press, 2003.

Hayden, Tom. "Gov. Jerry Brown: What Happens Here, Doesn't Stay Here." *Democracy Journal* (blog), September 9, 2014. http://tomhayden.com/home/gov-jerry-brown-what-happens-here-doesnt-stay-here.html.

Hecht, Sean. "Reflections on Environmental Justice and AB 32's Emissions Trading Program." *Legal Planet* (blog), March 23, 2011. http://legal-planet.org/2011/03/23reflections-on-environmental-justice-and-ab-32s-emissions-trading-program/.

Hess, Jeremy J., Josephine N. Malilay, and Alan J. Parkinson. "Climate Change: The Importance of Place." *American Journal of Preventive Medicine* 35, no. 5 (November 2008): 468–78.

Holifield, Ryan. "A Framework for a Critical Physical Geography of 'Sacrifice Zones': Physical Landscapes and Discursive Spaces of Frac Sand Mining in Western Wisconsin." *Geoforum* 85 (October 2017): 269–79.

———. "Neoliberalism and Environmental Justice Policy." In *Neoliberal Environments: False Promises and Unnatural Consequences*, edited by Nik Heynen, James McCarthy, Scott Prudham, and Paul Robbins, 202–14. London: Routledge, 2007.

Holifield, Ryan, Jayajit Chakraborty, and Gordon Walker, eds. *The Routledge Handbook of Environmental Justice*. London: Routledge, 2017.

Holmes, George. "Conservation's Friends in High Places: Neoliberalism, Networks, and the Transnational Conservation Elite." *Global Environmental Politics* 11, no. 4 (November 2011): 1–21.

Holmes-Gen, Bonnie. "Calif. Bill Could Generate Billions to Help Cope with Climate Change." *New America Media*, September 29, 2010. http://newamericamedia.org/2010/09/ab-1405-could-generate-billions-for-california.php.

Hooks, Gregory, and Chad L. Smith. "The Treadmill of Destruction: National Sacrifice Areas and Native Americans." *American Sociological Review* 69, no. 4 (August 2004): 558–75.

Hsia-Kiung, Katherine, Emily Reyna, and Timothy O'Connor. *Carbon Market California: A Comprehensive Analysis of the Golden State's Cap-and-Trade Program; Year One: 2012–2013*. EDF, 2014. http://www.edf.org/sites/default/files/content/ca-cap-and-trade_1yr_22_web.pdf.

Hull, R. Bruce. *Infinite Nature*. Chicago: University of Chicago Press, 2006.

Hulme, Mike. "Geographical Work at the Boundaries of Climate Change." *Transactions of the Institute of British Geographers* 33, no. 1 (January 2008): 5–11.

———. "Problems with Making and Governing Global Kinds of Knowledge." *Global Environmental Change* 20, no. 4 (October 2010): 558–64.

Hulme, Mike, and Martin Mahony. "Climate Change: What Do We Know about the IPCC?" *Progress in Physical Geography* 34, no. 5 (October 2010): 705–18.

ICLEI. *The Mitigation-Adaptation Connection: Milestones, Synergies and Contradictions*. 2009. http://itepsrv1.itep.nau.edu/itep_course_downloads/~GeneralAQInfo/Climate%20Change/ICLEIMitigationAdaptationConnection.pdf.

Iles, Alastair. "Identifying Environmental Health Risks in Consumer Products: Non-governmental Organizations and Civic Epistemologies." *Public Understanding of Science* 16, no. 4 (October 2007): 371–91.

Innes, Judith E. "Planning Theory's Emerging Paradigm: Communicative Action and Interactive Practice." *Journal of Planning Education and Research* 14, no. 3 (April 1995): 183–89.

Inside Cal/EPA. "Activists, Industry at Odds over California EPA's Cumulative Impacts Tool." August 31, 2012. https://insideepa.com/newsletters/inside-cal-epa.

———. "ARB Advisers Favor Carbon Revenues for Clean-Energy Projects." July 17, 2009. https://insideepa.com/newsletters/inside-cal-epa.

———. "California Community Fund Bill May Boost Press for GHG Auctions." September 2, 2010. https://insideepa.com/newsletters/inside-cal-epa.

———. "Carbon Revenue Bill Sets Stage for Future Policy, Budget Debates." October 9, 2009. https://insideepa.com/newsletters/inside-cal-epa.

———. "Contentious Bill to Earmark California GHG Funds for Poor Advances." June 30, 2011. https://insideepa.com/newsletters/inside-cal-epa.

IPCC. "Summary for Policymakers." In *Climate Change 2014: Synthesis Report*, edited by the Core Writing Team, Rajendra K. Pachauri, and Leo Mayer, 1–31. Geneva: IPCC, 2014. http://www.ipcc.ch/site/assets/uploads/2018/02/AR5_SYR_FINAL_SPM.pdf.

Irwin, Alan. *Citizen Science: A Study of People, Expertise and Sustainable Development*. London: Routledge, 1995.

Jasanoff, Sheila. "Cosmopolitan Knowledge: Climate Science and Global Civic

Epistemology." In *The Oxford Handbook of Climate Change and Society*, edited by John S. Dryzek, Richard B. Norgaard, and David Schlosberg, 129–43. Oxford: Oxford University Press, 2011.

———. *Designs on Nature: Science and Democracy in Europe and the United States*. Princeton, NJ: Princeton University Press, 2005.

———. *The Fifth Branch: Science Advisers as Policymakers*. Cambridge, MA: Harvard University Press, 1990.

———. "A New Climate for Society." *Theory, Culture & Society* 27, no. 2–3 (March 2010): 233–53.

———. "Ordering Knowledge, Ordering Society." In *States of Knowledge: The Co-production of Science and Social Order*, edited by Sheila Jasanoff, 13–45. London: Routledge, 2004.

Jasanoff, Sheila, and Marybeth Long Martello. "Conclusion: Knowledge and Governance." In *Earthly Politics: Local and Global in Environmental Governance*, edited by Sheila Jasanoff and Marybeth Long Martello, 335–50. Cambridge, MA: MIT Press, 2004.

Jessop, Bob. "Liberalism, Neoliberalism, and Urban Governance: A State-Theoretical Perspective." *Antipode* 34, no. 3 (July 2002): 452–72.

Johnson, Evan, ed. *California, Acre and Chiapas: Partnering to Reduce Emissions from Tropical Deforestation; Recommendations to Conserve Tropical Rainforests, Protect Local Communities and Reduce State-Wide Greenhouse Gas Emissions*. REDD Offset Working Group, 2013. https://www.arb.ca.gov/cc/capandtrade/sectorbased offsets/row-final-recommendations.pdf.

Johnson, Nathanael. "Ricardo Lara Grew Up in L.A.'s Dumping Grounds. Now He's Cleaning Them Up." *Grist*, December 4, 2017. https://grist.org/article /ricardo-lara-grew-up-in-l-a-s-dumping-grounds-now-hes-cleaning-them-up/.

Jonas, Andrew E. G. "The Scale Politics of Spatiality." *Environment and Planning D: Society and Space* 12, no. 3 (June 1994): 257–64.

Juhasz, Antonia. "Chevron's Refinery, Richmond's Peril." *Los Angeles Times*, August 14, 2012. http://articles.latimes.com/2012/aug/14/opinion/la-oe-0814-juhasz -chevron-refinery-pollution-20120814.

Juris, Jeffrey S. *Networking Futures: The Movements against Corporate Globalization*. Durham, NC: Duke University Press, 2008.

Kahn, Debra. "Despite Protests, Tropical States Move Forward on Incentives." *Climatewire*, September 12, 2018. https://www.eenews.net/stories/1060096637.

———. "Has a Decade of Golden State Climate Diplomacy Made a Difference?" *Climatewire*, September 23, 2016. https://www.eenews.net/stories/1060043294.

———. "What Schwarzenegger Learned about Climate Messaging." *Climatewire*, November 22, 2017. https://www.eenews.net/climatewire/stories/1060067171.

Kaijser, Anna, and Annica Kronsell. "Climate Change through the Lens of Intersectionality." *Environmental Politics* 23, no. 3 (2014): 417–33.

Karvonen, Andrew. *Politics of Urban Runoff: Nature, Technology, and the Sustainable City*. Cambridge, MA: MIT Press, 2011.

Kassel, Rich, and Joseph Annotti. *Dumping Dirty Diesels in Latin America: Reducing*

Black Carbon and Air Pollution from Diesel Engines in Latin American Countries. NRDC Report R:14-11-A. New York: NRDC, November 2014. https://assets.nrdc.org/sites/default/files/latin-america-diesel-pollution-report.pdf.

Kaswan, Alice. "A Broader Vision for Climate Policy: Lessons from California." *San Diego Journal of Climate & Energy Law* 9 (2018): 83–150.

———. "Climate Change and Environmental Justice: Lessons from the California Lawsuits." *San Diego Journal of Climate & Energy Law* 5 (2014): 1–42.

———. "Energy, Governance, and Market Mechanisms." *University of Miami Law Review* 72, no. 2 (2018): 476–579.

Kay, Jane, and Cheryl Katz. "Pollution, Poverty and People of Color: Living with Industry." *Scientific American*, June 4, 2012. https://www.scientificamerican.com/article/pollution-poverty-people-color-living-industry/.

Keck, Margaret E., and Kathryn Sikkink. *Activists beyond Borders: Advocacy Networks in International Politics*. Ithaca, NY: Cornell University Press, 1998.

Kelly, Alice B., and Nancy Lee Peluso. "Frontiers of Commodification: State Lands and Their Formalization." *Society & Natural Resources* 28, no. 5 (2015): 473–95.

King, Brian. *States of Disease: Political Environments and Human Health*. Oakland: University of California Press, 2017.

Kirsch, Emily. Interview by the author, October 2011.

Klein, Naomi. *This Changes Everything: Capitalism vs. the Climate*. New York: Simon & Schuster, 2014.

Knox-Hayes, Janelle. *The Cultures of Markets: The Political Economy of Climate Governance*. Oxford: Oxford University Press, 2016.

Krauss, Werner. "Localizing Climate Change: A Multi-sited Approach." In *Multi-sited Ethnography: Theory, Praxis and Locality in Contemporary Research*, edited by Mark-Anthony Falzon, 149–64. Farnham, UK: Ashgate, 2009.

Krieger, Nancy. "Embodiment: A Conceptual Glossary for Epidemiology." *Journal of Epidemiology & Community Health* 59, no. 5 (May 2005): 350–55.

———. "Embodiment and Ecosocial Theory: Interview with Nancy Krieger." By Kerstin Palm, Sigrid Schmitz, and Marion Mangelsdorf. *Freiburger Zeitschrift für Geschlechterstudien* 19, no. 2 (2013): 109–20.

———. *Embodying Inequality: Epidemiologic Perspectives*. London: Routledge, 2005.

———. "Theories for Social Epidemiology in the 21st Century: An Ecosocial Perspective." *International Journal of Epidemiology* 30, no. 4 (August 2001): 668–77.

Kunkel M., Catherine, and Daniel M. Kammen. "Design and Implementation of Carbon Cap and Dividend Policies." *Energy Policy* 39, no. 1 (January 2011): 477–86.

Kurtz, Hilda E. "Scale Frames and Counter-scale Frames: Constructing the Problem of Environmental Injustice." *Political Geography* 22, no. 8 (November 2003): 887–916.

Lang, Chris. "Protests in Chiapas against REDD: 'Stop the Land Grabs!'" *REDD-Monitor* (blog), September 28, 2012. http://www.redd-monitor.org/2012/09/28/protests-in-chiapas-against-redd-stop-the-land-grabs/.

Lansing, David, Rosemary-Claire Collard, Jessica Dempsey, Juanita Sundberg,

Nik Heynen, Bram Büscher, Wolfram Dressler, and Robert Fletcher. Review symposium on *Nature™ Inc.: Environmental Conservation in a Neoliberal Age*, edited by Bram Büscher, Wolfram Dressler, and Robert Fletcher. *Environment and Planning A: Economy and Space* 47, no. 11 (November 2015): 2389–408.

Latino Caucus member. Interview by the author, August 2013.

———. Interview by the author, August 2017.

———. Interview by the author, November 2017.

Latour, Bruno. "Drawing Things Together." In *Representation in Scientific Practice*, edited by Michael Lynch and Steve Woolgar, 19–68. Cambridge, MA: MIT Press, 1990.

———. *Science in Action: How to Follow Scientists and Engineers through Society*. Cambridge, MA: Harvard University Press, 1987.

Lebel, Louis, Po Garden, and Masao Imamura. "The Politics of Scale, Position, and Place in the Governance of Water Resources in the Mekong Region." *Ecology and Society* 10, no. 2 (December 2005). https://www.ecologyandsociety.org/vol10/iss2/art18/.

Lejano, Raul P., and Rei Hirose. "Testing the Assumptions behind Emissions Trading in Non-market Goods: The RECLAIM Program in Southern California." *Environmental Science & Policy* 8, no. 4 (August 2005): 367–77.

Lemos, Maria Carmen, and Barbara J. Morehouse. "The Co-production of Science and Policy in Integrated Climate Assessments." *Global Environmental Change* 15, no. 1 (April 2005): 57–68.

Lerner, Steve. *Sacrifice Zones: The Front Lines of Toxic Chemical Exposure in the United States*. Cambridge, MA: MIT Press, 2010.

Lerza, Catherine. *A Perfect Storm: Lessons from the Defeat of Proposition 23*. Funders Network for Transforming the Global Economy, September 2011. https://edgefunders.org/wp-content/uploads/2015/09/Prop23CaseStudy_000.pdf.

Leung, Wendy. "Residents Protest Proposed Railyard Air Pollution Plan: More than 100 Residents from Rail Adjacent Communities Protested." *San Bernardino Sun*, June 20, 2010.

Li, Jennifer Chung-I. "Including the Feedback of Local Health Improvement in Assessing Costs and Benefits of GHG Reduction." *Review of Urban & Regional Development Studies* 14, no. 3 (November 2002): 282–304.

Liévanos, Raoul S. "Certainty, Fairness, and Balance: State Resonance and Environmental Justice Policy Implementation." *Sociological Forum* 27, no. 2 (June 2012): 481–503.

———. "Retooling CalEnviroScreen: Cumulative Pollution Burden and Race-Based Environmental Health Vulnerabilities in California." *International Journal of Environmental Research and Public Health* 15, no. 4 (April 2018): 762–87.

Liévanos, Raoul S., Jonathan K. London, and Julie Sze. "Uneven Transformations and Environmental Justice: Regulatory Science, Street Science, and Pesticide Regulation in California." In *Technoscience and Environmental Justice: Expert Cultures in a Grassroots Movement*, edited by Gwen Ottinger and Benjamin Cohen, 201–28. Cambridge, MA: MIT Press, 2011.

Lim, Merlyna. "A Cyber-Urban Space Odyssey: The Spatiality of Contemporary Social Movements." In *Geographies of Information*, edited by Ali Fard and Taraneh Meshkani, 117–23. New Geographies 7. Cambridge, MA: Harvard University Graduate School of Design, 2015.

Lin, Serena W. *Understanding Climate Change: An Equitable Framework*. Oakland, CA: PolicyLink, 2008.

Lind, Michael. "The Five Worldviews that Define American Politics." *Salon*, January 12, 2011. https://www.salon.com/2011/01/12/lind_five_worldviews/.

Liverman, Diana M., and Silvina Vilas. "Neoliberalism and the Environment in Latin America." *Annual Review of Environment and Resources* 31 (2006): 327–63.

Locatelli, Bruno. *Synergies between Adaptation and Mitigation in a Nutshell*. Center for International Forestry Research, August 2011. https://www.cifor.org/fileadmin/fileupload/cobam/ENGLISH-Definitions%26ConceptualFramework.pdf.

Lohmann, Larry. "Performative Equations and Neoliberal Commodification: The Case of Climate." In *Nature™ Inc.: Environmental Conservation in the Neoliberal Age*, edited by Bram Büscher, Wolfram Dressler, and Robert Fletcher, 158–82. Tucson: University of Arizona Press, 2014.

London, Jonathan, Alex Karner, Julie Sze, Dana Rowan, Gerardo Gambirazzio, and Deb Niemeier. "Racing Climate Change: Collaboration and Conflict in California's Global Climate Change Policy Arena." *Global Environmental Change* 23, no. 4 (August 2013): 791–99.

London, Jonathan K., Julie Sze, and Raoul S. Liévanos. "Problems, Promise, Progress, and Perils: Critical Reflections on Environmental Justice Policy Implementation in California." *UCLA Journal of Environmental Law and Policy* 26, no. 2 (2008): 255–89.

Long, Douglas G. "'Utility' and the 'Utility Principle': Hume, Smith, Bentham, Mill." *Utilitas* 2, no. 1 (May 1990): 12–39.

Lopez, Pia. "Crossing Borders: California Tries to Cultivate Green Roots with Chiapas." *Sacramento Bee*, March 10, 2013.

Lueders, Jesse, Cara Horowitz, Ann E. Carlson, Sean B. Hecht, and Edward A. Parson. *The California REDD+ Experience: The Ongoing Political History of California's Initiative to Include Jurisdictional REDD+ Offsets within Its Cap-and-Trade System*. Emmett Institute on Climate Change and the Environment at the UCLA School of Law, November 19, 2014.

Lusvardi, Wayne. "CA Air Board May Invalidate 1.3 Million Pollution-Offset Credits." *CalWatchdog* (blog), June 9, 2014. https://calwatchdog.com/2014/06/09/ca-air-board-may-invalidate-1-3-million-pollution-offset-credits/.

MacDonald, Kenneth Iain. "The Devil Is in the (Bio)diversity: Private Sector 'Engagement' and the Restructuring of Biodiversity Conservation." *Antipode* 42, no. 3 (June 2010): 513–50.

Maibach, Edward W., Matthew Nisbet, Paula Baldwin, Karen Akerlof, and Guoqing Diao. "Reframing Climate Change as a Public Health Issue: An Exploratory Study of Public Reactions." *BMC Public Health* 10 (2010): 1–11.

Marcus, George E. "Ethnography in/of the World System: The Emergence of Multi-sited Ethnography." *Annual Review of Anthropology* 24 (1995): 95–117.

Marcus, George E., and Michael M. J. Fischer. *Anthropology as Cultural Critique: An Experimental Moment in the Human Sciences*. Chicago: University of Chicago Press, 1986.

Market Advisory Committee. *Recommendations for Designing a Greenhouse Gas Cap-and-Trade System for California: Recommendations of the Market Advisory Committee to the California Air Resources Board*. June 30, 2007. http://www.energy.ca.gov/2007publications/ARB-1000-2007-007/ARB-1000-2007-007.PDF.

Martello, Marybeth Long. "Arctic Indigenous Peoples as Representations and Representatives of Climate Change." *Social Studies of Science* 38, no. 3 (June 2008): 351–76.

———. "A Paradox of Virtue? 'Other' Knowledges and Environment-Development Politics." *Global Environmental Politics* 1, no. 3 (August 2001): 114–41.

Martello, Marybeth Long, and Sheila Jasanoff. "Introduction: Globalization and Environmental Governance." In *Earthly Politics: Local and Global in Environmental Governance*, edited by Sheila Jasanoff and Marybeth Long Martello, 1–29. Cambridge, MA: MIT Press, 2004.

Martin, Mark. "Sacramento: Núñez Slams Governor on Emission Law." *SFGate*, October 17, 2006. https://www.sfgate.com/green/article/SACRAMENTO-N-ez-slams-governor-on-emission-2485726.php.

Martinez-Alier, Joan, Giuseppe Munda, and John O'Neill. "Weak Comparability of Values as a Foundation for Ecological Economics." *Ecological Economics* 26, no. 3 (September 1998): 277–86.

Massey, Doreen B., ed. *Power-Geometries and the Politics of Space-Time: Hettner-Lecture 1998*. Heidelberg, Germany: Department of Geography, University of Heidelberg, 1999.

Mayer, Steven. "Red Alert: The Color Purple Declared as Poor Air Impacts Schools." *Bakersfield Californian*, January 13, 2014. https://www.bakersfield.com/news/red-alert-the-color-purple-declared-as-poor-air-impacts/article_f0d56c63-2f79-54ad-a449-7e67907708a9.html.

Mazmanian, Daniel A., John Jurewitz, and Hal Nelson. "California's Climate Change Policy: The Case of a Subnational State Actor Tackling a Global Challenge." *Journal of Environment & Development* 17, no. 4 (December 2008): 401–23.

McAdam, Doug, John D. McCarthy, and Mayer N. Zald. "Introduction: Opportunities, Mobilizing Structures, and Framing Processes—Toward a Synthetic, Comparative Perspective on Social Movements." In *Comparative Perspectives on Social Movements: Political Opportunities, Mobilizing Structures, and Cultural Framings*, edited by Doug McAdam, John D. McCarthy, and Mayer N. Zald, 1–20. New York: Cambridge University Press, 1996.

McAfee, Kathleen. "Selling Nature to Save It? Biodiversity and Green Developmentalism." *Environment and Planning D: Society and Space* 17, no. 2 (April 1999): 133–54.

McDonnell, Tim. "Can California Help the Paris Climate Talks Succeed?" *Mother Jones*, December 8, 2015. https://www.motherjones.com/environment/2015/12/california-climate-change-cop21-paris-tom-steyer/.

McFarlane, Colin. "Translocal Assemblages: Space, Power and Social Movements." *Geoforum* 40, no. 4 (July 2009): 561–67.

McGreevy, Patrick. "He Rallied Support for California's Climate Change Fight. Now Chad Mayes Is Out as Assembly Republican Leader." *Los Angeles Times*, August 24, 2017. https://www.latimes.com/politics/la-pol-ca-republicans-assembly-leader-dahle-20170824-story.html.

Meidinger, Errol. "Regulatory Culture: A Theoretical Outline." *Law & Policy* 9, no. 4 (October 1987): 355–86.

Melendez, Lyanne. "Politicians, Celebs Talk Climate Change at Global Climate Action Summit amid Protests in San Francisco." *ABC 7 News*, September 13, 2018. https://abc7news.com/science/politicians-celebs-talk-climate-change-amid-protests-in-sf/4236426/.

Mendez, Michael. "Assessing Local Climate Action Plans for Public Health Co-benefits in Environmental Justice Communities." *Local Environment* 20, no. 6 (June 2015): 637–63.

———. "From the Street: Civic Epistemologies of Urban Climate Change." In *Spatializing Politics: Essays on Power and Place*, edited by Delia Duong Ba Wendel and Fallon Samuels Aidoo, 337–65. Cambridge, MA: Harvard University Graduate School of Design, 2016.

Mikati, Ihab, Adam F. Benson, Thomas J. Luben, Jason D. Sacks, and Jennifer Richmond-Bryant. "Disparities in Distribution of Particulate Matter Emission Sources by Race and Poverty Status." *American Journal of Public Health* 108, no. 4 (April 2018): 480–85.

Miller, Byron. "Political Empowerment, Local-Central State Relations, and Geographically Shifting Political Opportunity Structures: Strategies of the Cambridge, Massachusetts, Peace Movement." *Political Geography* 13, no. 5 (September 1994): 393–406.

Miller, Clark A. "Civic Epistemologies: Constituting Knowledge and Order in Political Communities." *Sociology Compass* 2, no. 6 (November 2008): 1896–919.

———. "New Civic Epistemologies of Quantification: Making Sense of Indicators of Local and Global Sustainability." *Science, Technology, & Human Values* 30, no. 3 (Summer 2005): 403–32.

Minkler, Meredith, and Nina Wallerstein, eds. *Community-Based Participatory Research for Health: From Process to Outcomes*. 2nd ed. San Francisco: Jossey-Bass, 2008.

Mitlin, Diana, and David Satterthwaite. *Urban Poverty in the Global South: Scale and Nature*. London: Routledge, 2013.

Mohai, Paul, David Pellow, and J. Timmons Roberts. "Environmental Justice." *Annual Review of Environment and Resources* 34 (2009): 405–30.

Montañez, Cindy. Interview by the author, April 2006.

———. Interview by the author, December 2017.

Moore, Shirley Ann Wilson. *To Place Our Deeds: The African American Community in Richmond, California, 1910–1963*. Berkeley: University of California Press, 2000.

Morain, Dan. "Cool Bid for State's Pollution Program Is Lost on Arkansas Incinerator." *Sacramento Bee*, June 28, 2014. https://www.sacbee.com/opinion/opn-columns-blogs/dan-morain/article2602431.html.

Morello-Frosch, Rachel, and Manuel Pastor. "Environmental Justice and Vulnerable Populations." In *Environmental Health: From Global to Local*, edited by Howard Frumkin, 251–72. 3rd ed. San Francisco: Jossey-Bass, 2016.

Morgen, Sandra. *Into Our Own Hands: The Women's Health Movement in the United States, 1969–1990*. New Brunswick, NJ: Rutgers University Press, 2002.

Morris, Jim. "'The Fear of Dying' Pervades Southern California's Oil-Polluted Enclaves." Center for Public Integrity, October 30, 2017. https://www.publicintegrity.org/2017/10/30/21221/fear-dying-pervades-southern-californias-oil-polluted-enclaves.

Moser, Susanne C., and Lisa Dilling, eds. *Creating a Climate for Change: Communicating Climate Change and Facilitating Social Change*. Cambridge: Cambridge University Press, 2007.

Mulady, Kathy. "Richmond Refinery Fire Unites Communities Divided." *Equal Voice News* (blog), October 2, 2012. https://www.equalvoiceforfamilies.org/richmond-refinery-fire-unites-communities-divided/.

Nachbaur, James, and Tiffany Roberts. *Evaluating the Policy Trade-Offs in ARB's Cap-and-Trade Program*. Legislative Analyst's Office, February 9, 2012. https://lao.ca.gov/reports/2012/rsrc/cap-and-trade/cap-and-trade-020912.pdf.

National Research Council. *Adaptive Management for Water Resources Project Planning*. Washington, DC: National Academies Press, 2004. https://doi.org/10.17226/10972.

Naughton-Treves, Lisa, and Cathy Day, eds. *Lessons about Land Tenure, Forest Governance and REDD+: Case Studies from Africa, Asia and Latin America*. Madison: University of Wisconsin-Madison Land Tenure Center, January 2012. https://www.nelson.wisc.edu/ltc/docs/Lessons-about-Land-Tenure-Forest-Governance-and-REDD.pdf.

Naylor, Brian. "White House Reverses Effort to Delay Obama Ozone Regulations." *All Things Considered*, NPR, August 3, 2017. https://www.npr.org/2017/08/03/541432494/white-house-reverses-effort-to-delay-obama-ozone-regulations.

Nemet, G. F., T. Holloway, and P. Meier. "Implications of Incorporating Air-Quality Co-benefits into Climate Change Policymaking." *Environmental Research Letters* 5, no. 1 (January–March 2010): 1–9.

New Forests and the Yurok Tribe. "New Forests and Yurok Tribe Achieve Regulatory Approval of First Compliance Forest Carbon Offset Project for California Carbon Market." Press release, April 10, 2014. https://www.newforests.com.au/wp-content/uploads/2014/09/20140410-New-Forests-Yurok-Tribe-announce-regulatory-project-appoval.pdf.

Newtown, Mark. *LAO's Critique of the AB 32 Scoping Plan Economic Analysis*. Report presented to the Assembly Natural Resources Committee, March 9, 2009.

Nicholls, Walter, Byron Miller, and Justin Beaumont, eds. *Spaces of Contention: Spatialities and Social Movements*. London: Routledge, 2013.

Niemeier, Deb, Thomas D. Beamish, Alissa Kendall, Ryken Grattet, Jonathan London, Carolyn de la Pena, and Julie Sze. "Characterizing the Impacts of Uncertainty in the Policy Process: Climate Science, Policy Construction, and Local Governance Decisions." In *Transition towards Sustainable Mobility: The Role of Instruments, Individuals and Institutions*, edited by Harry Geerlings, Yoram Shiftan, and Dominic Stead, 119–36. Farnham, UK: Ashgate, 2012.

No REDD Tour. "Indigenous Peoples Respond to False Climate Change Solutions!" September 13, 2012. https://noreddtour.wordpress.com/.

Norgaard, Kari Marie. *Living in Denial: Climate Change, Emotions, and Everyday Life*. Cambridge, MA: MIT Press, 2011.

NRDC and E2. "Making a Difference (E2 and AB 32)." October 2006. https://members.e2.org/ext/doc/MakingADifference%20_E2andAB32.pdf.

Núñez, Fabian. "Cap-and-Trade: Investing in California's Future." *Sacramento Bee*, September 27, 2012.

———. "Connecting the Dots between Global Warming Bill and Health of Those Exposed to Carbon-Based Gases." *California Progress Report*, September 11, 2006. https://www.frankejames.com/connecting-the-dots-between-global-warming-bill-and-health-of-those-exposed-to-carbon-based-gases/.

O'Brien, Claudia, and Bart Kempf. "CARB Invalidates Offsets for Facility's Alleged RCRA Noncompliance." Latham's Clean Energy Law Report, November 19, 2014. https://www.cleanenergylawreport.com/environmental-and-approvals/carb-invalidates-offsets-for-facilitys-alleged-rcra-noncompliance/.

O'Brien, Luke. "California Official Says Schwarzenegger Misled on Global Warming." *Wired*, July 3, 2007. https://www.wired.com/2007/07/california-offi/.

OCAC. Media content distributed at public hearings and community events. 2009.

OCAC member. Interview by the author, May 2011.

———. Interview by the author, May 2013.

OCAC Steering Committee member. Interview by the author, May 2013.

———. Interview by the author, August 2013.

O'Connor, Tim. Testimony, Assembly Natural Resources Committee hearing, Sacramento, California, August 13, 2013.

ODPHP. "Social Determinants of Health." US Department of Health and Human Services, 2014. https://www.healthypeople.gov/2020/topics-objectives/topic/social-determinants-of-health.

OECD. "Command-and-Control Policy." Glossary of Statistical Terms. 2001. http://stats.oecd.org/glossary/detail.asp?ID=383.

OEHHA. "Call for Applications to the Cumulative Impacts and Precautionary Approaches Work Group." 2007. https://oehha.ca.gov/media/downloads/calenviroscreen/document/cipaworkgroupsolicitation.pdf.

———. *Draft California Communities Environmental Health Screening Tool (CalEnviro-*

Screen): Proposed Methods & Indicators. July 30, 2012. https://oehha.ca.gov/media/downloads/calenviroscreen/report/draftcalenviroscreen073012.pdf.

Okereke, Chukwumerije, and Kate Dooley. "Principles of Justice in Proposals and Policy Approaches to Avoided Deforestation: Towards a Post-Kyoto Climate Agreement." *Global Environmental Change* 20, no. 1 (February 2010): 82–95.

Omi, Michael, and Howard Winant. *Racial Formation in the United States*. 3rd ed. New York: Routledge, 2015.

Osborne, Tracey. "Fixing Carbon, Losing Ground: Payments for Environmental Services and Land (In)security in Mexico." *Human Geography* 6, no. 1 (2013): 119–33.

———. "Tradeoffs in Carbon Commodification: A Political Ecology of Common Property Forest Governance." *Geoforum* 67 (December 2015): 64–77.

Osofsky, Hari M. "The Intersection of Scale, Science, and Law in *Massachusetts v. EPA*." In *Adjudicating Climate Change: State, National, and International Approaches*, edited by William C. G. Burns and Hari M. Osofsky, 129–44. Cambridge: Cambridge University Press, 2009.

Ostrander, Madeline. "Will California's Cap and Trade Be Fair?" *Nation*, March 20, 2013. https://www.thenation.com/article/will-californias-cap-and-trade-be-fair/.

Palmer, Gary B. *Toward a Theory of Cultural Linguistics*. Austin: University of Texas Press, 1996.

Parino, Sofia. Interview by the author, July 2013.

Parino, Sofia L., Adrienne Block, Strela Cervas, Maria S. Covarrubias, Vianey Nunez, Tony Perez, Jesse N. Marquez, et al. "Comments on the Supplement to the AB 32 Scoping Plan Functional Equivalent Document." July 28, 2011. https://www.arb.ca.gov/lists/ceqa-sp11/119-commentsandexhibits.pdf.

Park, Angela. *Everybody's Movement: Environmental Justice and Climate Change*. Washington, DC: Environmental Support Center, December 2009. https://kresge.org/sites/default/files/Everybodys-movement-climate-social-justice.pdf.

Pastor, Manuel. "Environmental Justice: Reflections from the United States." Political Economy Research Institute and Centre for Science and Environment, November 2002. Paper presented at the International Conference on Natural Assets, Tagaytay City, the Philippines, January 8–11, 2003. https://www.peri.umass.edu/publication/item/140-environmental-justice-reflections-from-the-united-states-2002.

———. *State of Resistance: What California's Dizzying Descent and Remarkable Resurgence Mean for America's Future*. New York: New Press, 2018.

Pastor, Manuel, Rachel Morello-Frosch, James Sadd, and Justin Scoggins. "Risky Business: Cap-and-Trade, Public Health, and Environmental Justice." In *Urbanization and Sustainability: Linking Urban Ecology, Environmental Justice and Global Environmental Change*, edited by Christopher G. Boone and Michail Fragkias, 75–94. Dordrecht, the Netherlands: Springer, 2013.

Peace, Janet, and Robert N. Stavins. *Meaningful and Cost Effective Climate Policy: The*

Case for Cap and Trade. Arlington, VA: Pew Center on Global Climate Change, June 2010. https://www.c2es.org/document/meaningful-and-cost-effective-climate-policy-the-case-for-cap-and-trade/.

Pelling, Mark. *Adaptation to Climate Change: From Resilience to Transformation*. London: Routledge, 2010.

Pellow, David Naguib. "Politics by Other Greens: The Importance of Transnational Environmental Justice Movement Networks." In *Environmental Inequalities beyond Borders: Local Perspectives on Global Injustices*, edited by JoAnn Carmin and Julian Agyeman, 247–66. Cambridge, MA: MIT Press, 2011.

———. *Resisting Global Toxics: Transnational Movements for Environmental Justice*. Cambridge, MA: MIT Press, 2007.

———. "Toward a Critical Environmental Justice Studies: Black Lives Matter as an Environmental Justice Challenge." *Du Bois Review: Social Science Research on Race* 13, no. 2 (Fall 2016): 221–36.

———. *What Is Critical Environmental Justice?* Cambridge: Polity, 2018.

Peth, Simon Alexander. "What Is Translocality? A Refined Understanding of Place and Space in a Globalized World." *Connecting the Spots: Notes on Migration and Environment from a Geographical Perspective* (blog), November 9, 2014. http://transre.uni-bonn.de/en/blog/what-translocality/.

Picq, Manuela. "Will California Fall into the REDD Trap?" *Al Jazeera*, May 7, 2013. https://www.aljazeera.com/indepth/opinion/2013/05/20135613232989660.html.

Pittel, Karen, and Dirk T. G. Rübbelke. "Climate Policy and Ancillary Benefits: A Survey and Integration into the Modelling of International Negotiations on Climate Change." *Ecological Economics* 68, no. 1–2 (December 2008): 210–20.

Porta, Donatella della, and Mario Diani. *Social Movements: An Introduction*. 2nd ed. Malden, MA: Wiley-Blackwell, 2006.

Pulido, Laura. *Environmentalism and Economic Justice: Two Chicano Struggles in the Southwest*. Tucson: University of Arizona Press, 1996.

Pulido, Laura, Ellen Kohl, and Nicole-Marie Cotton. "State Regulation and Environmental Justice: The Need for Strategy Reassessment." *Capitalism Nature Socialism* 27, no. 2 (April 2016): 12–31.

Rabe, Barry. "Governing the Climate from Sacramento." In *Unlocking the Power of Networks: Keys to High-Performance Government*, edited by Stephen Goldsmith and Donald F. Kettl, 34–61. Cambridge, MA: Ash Institute for Democratic Governance and Innovation; Washington, DC: Brookings Institution Press, 2009.

Rabin, Yale. "Expulsive Zoning: The Inequitable Legacy of Euclid." In *Zoning and the American Dream: Promises Still to Keep*, edited by Charles M. Haar and Jerold S. Kayden, 101–21. Chicago: Planners Press, 1989.

Ramo, Alan. "Comments on the REDD Offset Working Group's Draft Recommendations on Partnering to Reduce Emissions from Tropical Deforestation—California, Acre and Chiapas." April 22, 2013.

Ransby, Barbara. *Ella Baker and the Black Freedom Movement: A Radical Democratic Vision*. Chapel Hill: University of North Carolina Press, 2003.

Richter, Burton. *Beyond Smoke and Mirrors: Climate Change and Energy in the 21st Century*. Cambridge: Cambridge University Press, 2010.

Rodriquez, Matt. Comments during Senate Rules Committee confirmation hearing, Sacramento, California, February 12, 2012.

Rogers, Paul. "California's Global Warming Law Takes a Hit." *Mercury News*, March 21, 2011. https://www.mercurynews.com/2011/03/21/californias-global-warming-law-takes-a-hit/.

Roland-Holst, David. "Economic Growth and Greenhouse Gas Mitigation in California." August 2006. http://www.energy.ca.gov/2007publications/UCB-1000-2007-008/UCB-1000-2007-008-ES.PDF.

Rosenberg, Paul. "Gov. Schwarzenegger's Global Warming Act Called Hot Air; Governor Fires Sawyer Then Denies It." ArnoldWatch, July 27, 2007. http://www.arnoldwatch.org/articles/articles_001067.php3.

Rosenhall, Laurel. "Mercury Lobbying Firm Working to Open Office in Mexico." *Sacramento Bee*, August 7, 2014. http://www.sacbee.com/news/politics-government/capitol-alert/article2606045.html.

———. "Three Fined for Covert Lobbying in California." *Sacramento Bee*, September 10, 2013. http://www.sacbee.com/news/investigations/lobbying-influence/article6692601.html.

Rosenthal, Joyce Klein, and Dana Brechwald. "Climate Adaptive Planning for Preventing Heat-Related Health Impacts in New York City." In *Climate Change Governance*, edited by Jörg Knieling and Walter Leal Filho, 205–25. Berlin: Springer, 2013.

Rosenzweig, Cynthia, William Solecki, Stephen A. Hammer, and Shagun Mehrotra. "Cities Lead the Way in Climate-Change Action." *Nature* 467, no. 7318 (October 21, 2010): 909–11.

Roth, Sammy. "At Paris Climate Talks, Nations Looking to California." *Desert Sun*, November 14, 2015. https://www.desertsun.com/story/news/environment/2015/11/14/paris-climate-talks-nations-look-california/75540806/.

Sadd, James L., Manuel Pastor, Rachel Morello-Frosch, Justin Scoggins, and Bill Jesdale. "Playing It Safe: Assessing Cumulative Impact and Social Vulnerability through an Environmental Justice Screening Method in the South Coast Air Basin, California." *International Journal of Environmental Research and Public Health* 8, no. 5 (May 2011): 1441–59.

San Joaquin Valley Air Pollution Control District. *Presidential Transition White Paper*. November 2016. http://www.valleyair.org/documents/PRESIDENTIAL-TRANSITION-WHITE-PAPER.pdf.

Sasser, Jade S. "Population, Climate Change, and the Embodiment of Environmental Crisis." In *Systemic Crises of Global Climate Change: Intersections of Race, Class, and Gender*, edited by Phoebe Godfrey and Denise Torres, 57–70. London: Routledge, 2016.

Sayre, Nathan F. "Ecological and Geographical Scale: Parallels and Potential for Integration." *Progress in Human Geography* 29, no. 3 (June 2005): 276–90.

Scarre, Geoffrey. *Utilitarianism*. London: Routledge, 1996.

Schaeffer, Eric, Tom Pelton, and Keene Kelderman. *Paying Less to Pollute: A Year of Environmental Enforcement under the Trump Administration*. Environmental Integrity Project, February 15, 2018. http://www.environmentalintegrity.org/wp-content/uploads/2017/02/Enforcement-Report.pdf.

Scheper-Hughes, Nancy, and Margaret M. Lock. "The Mindful Body: A Prolegomenon to Future Work in Medical Anthropology." *Medical Anthropology Quarterly* 1, no. 1 (March 1987): 6–41.

Schlosberg, David. *Defining Environmental Justice: Theories, Movements, and Nature*. Oxford: Oxford University Press, 2007.

———. "Reconceiving Environmental Justice: Global Movements and Political Theories." *Environmental Politics* 13, no. 3 (September 2004): 517–40.

Schlosberg, David, and David Carruthers. "Indigenous Struggles, Environmental Justice, and Community Capabilities." *Global Environmental Politics* 10, no. 4 (November 2010): 12–35.

Schlosberg, David, and Lisette B. Collins. "From Environmental to Climate Justice: Climate Change and the Discourse of Environmental Justice." *WIREs Climate Change* 5, no. 3 (May/June 2014): 359–74.

Schofield, Philip. *Utility and Democracy: The Political Thought of Jeremy Bentham*. Oxford: Oxford University Press, 2006.

Schroeder, Richard A. "Geographies of Environmental Intervention in Africa." *Progress in Human Geography* 23, no. 3 (September 1999): 359–78.

Schwarzenegger, Arnold. "Executive Order S-20-06." State of California, Office of the Governor, 2006. http://www.climatechange.ca.gov/state/executive_orders.html.

———. "Governor's Veto Message." California Legislative Information, September 30, 2010. https://leginfo.legislature.ca.gov/faces/billStatusClient.xhtml?bill_id=200920100AB1405.

———. "Memorandum of Understanding on Environmental Cooperation between the State of Acre of the Federative Republic of Brazil, the State of Chiapas of the United Mexican States, and the State of California of the United States of America." November 2010.

Schwebs, Monica A. "Global Warming I: Developing Cap and Trade Programs to Reduce Greenhouse Gas Emissions." *Environmental Law News* 16 (2007): 34–43.

Scott, James C. *Seeing like a State: How Certain Schemes to Improve the Human Condition Have Failed*. New Haven, CT: Yale University Press, 1998.

Sen, Amartya. "Capability and Well-Being." In *The Quality of Life*, edited by Martha C. Nussbaum and Amartya Sen, 30–53. Oxford: Clarendon Press, 1993.

Senior capitol staff member. Interview by the author, August 2013.

———. Interview by the author, August 2017.

———. Interview by the author, June 2013.

———. Interview by the author, May 2013.

———. Interview by the author, October 2013.

Senior Schwarzenegger appointee. Interview by the author, August 2013.

Shah, Bindi V. *Laotian Daughters: Working toward Community, Belonging, and Environmental Justice*. Philadelphia: Temple University Press, 2012.

Sherwin, Murray, Sally Davenport, and Graham Scott. *Regulatory Institutions and Practices*. New Zealand Productivity Commission, June 2014. https://www.productivity.govt.nz/sites/default/files/regulatory-institutions-and-practices-final-report.pdf.

Shonkoff, Seth B., Rachel Morello-Frosch, Manuel Pastor, and James Sadd. "The Climate Gap: Environmental Health and Equity Implications of Climate Change and Mitigation Policies in California—A Review of the Literature." *Climatic Change* 109, no. S1 (December 2011): 485–503.

———. "Minding the Climate Gap: Environmental Health and Equity Implications of Climate Change Mitigation Policies in California." *Environmental Justice* 2, no. 4 (December 2009): 173–77.

Silverstein, Ken. "New Ozone Pollution Rules Take Effect over Objections from Trump's Team." *Forbes*, October 1, 2017. https://www.forbes.com/sites/kensilverstein/2017/10/01/new-ozone-pollution-rules-take-effect-over-objections-from-trumps-team/.

Skocpol, Theda. "Naming the Problem: What It Will Take to Counter Extremism and Engage Americans in the Fight against Global Warming." Unpublished paper, January 2013. Prepared for the symposium on the Politics of America's Fight against Global Warming, Harvard University, Cambridge, MA, February 14, 2013. https://climateaccess.org/system/files/Skocpol_Naming%20the%20Problem.pdf.

Skorupski, John. *John Stuart Mill*. London: Routledge, 1989.

Smith, Kirk R., Alistair Woodward, Diarmid Campbell-Lendrum, Dave D. Chadee, Yasushi Honda, Qiyong Liu, Jane M. Olwoch, Boris Revich, and Rainer Sauerborn. "Human Health: Impacts, Adaptation, and Co-benefits." In *Climate Change 2014: Impacts, Adaptation, and Vulnerability; Part A: Global and Sectoral Aspects. Contribution of Working Group II to the Fifth Assessment Report of the Intergovernmental Panel on Climate Change*, edited by Christopher B. Field, Vicente R. Barros, David Jon Dokken, Katharine J. Mach, Michael D. Mastrandrea, T. Eren Bilir, Monalisa Chatterjee, et al., 704–54. Cambridge: Cambridge University Press, 2014. https://www.ipcc.ch/site/assets/uploads/2018/02/WGIIAR5-Chap11_FINAL.pdf.

Smith, Scott. "California Farm Region Plagued by Dirty Air Looks to Trump." *Phys.org*, July 20, 2017. https://phys.org/news/2017-07-california-farm-region-plagued-dirty.html.

Somers, Margaret R., and Fred Block. "From Poverty to Perversity: Ideas, Markets, and Institutions over 200 Years of Welfare Debate." *American Sociological Review* 70, no. 2 (April 2005): 260–87.

Soros, George. *The Crisis of Global Capitalism: Open Society Endangered*. New York: Little, Brown, 2000.

Stavins, Robert. "Why the Environmental Justice Lawsuit against California's Climate Law Is Misguided." *Grist*, May 23, 2011. http://grist.org/climate-policy/2011-05-23-environmental-justice-lawsuit-against-californias-climate-law/.

Stehr, Nico, and Hans von Storch. "Introduction to Papers on Mitigation and Adaptation Strategies for Climate Change: Protecting Nature from Society or Protecting Society from Nature?" *Environmental Science & Policy* 8, no. 6 (December 2005): 537–40.

Stockton, Nick. "California Marches into Paris to Fight the Climate Apocalypse." *Wired*, December 8, 2015. https://www.wired.com/2015/12/california-marches-into-paris-to-fight-the-climate-apocalypse/.

Stone, Daniel. "California Tackles Climate Change, but Will Others Follow?" *National Geographic News*, November 16, 2012. https://news.nationalgeographic.com/news/energy/2012/11/121116-california-cap-and-trade/.

Sunderlin, W. D., E. O. Sills, A. E. Duchelle, A. D. Ekaputri, D. Kweka, A. Toniolo, S. Ball, et al. "REDD+ at a Critical Juncture: Assessing the Limits of Polycentric Governance for Achieving Climate Change Mitigation." *International Forestry Review* 17, no. 4 (2015): 1–14.

Swyngedouw, Erik, and Nikolas C. Heynen. "Urban Political Ecology, Justice and the Politics of Scale." *Antipode* 35, no. 5 (November 2003): 898–918.

Szasz, Andrew, and Michael Meuser. "Environmental Inequalities: Literature Review and Proposals for New Directions in Research and Theory." *Current Sociology* 45, no. 3 (July 1997): 99–120.

Sze, Julie. *Noxious New York: The Racial Politics of Urban Health and Environmental Justice*. Cambridge, MA: MIT Press, 2007.

Sze, Julie, Gerardo Gambirazzio, Alex Karner, Dana Rowan, Jonathan London, and Deb Niemeier. "Best in Show? Climate and Environmental Justice Policy in California." *Environmental Justice* 2, no. 4 (December 2009): 179–84.

Sze, Julie, and Jonathan K. London. "Environmental Justice at the Crossroads." *Sociology Compass* 2, no. 4 (July 2008): 1331–54.

Tarrow, Sidney. *Power in Movement: Social Movements, Collective Action and Politics*. Cambridge: Cambridge University Press, 1994.

Taruc, Mari Rose. Interview by the author, August 2017.

———. Interview by the author, May 2013.

Taylor, Dorceta E. *The Rise of the American Conservation Movement: Power, Privilege, and Environmental Protection*. Durham, NC: Duke University Press, 2016.

———. *Toxic Communities: Environmental Racism, Industrial Pollution, and Residential Mobility*. New York: New York University Press, 2014.

Tesh, Sylvia N., and Bruce A. Williams. "Identity Politics, Disinterested Politics, and Environmental Justice." *Polity* 28, no. 3 (Spring 1996): 285–305.

Thompson, Tammy M., Sebastian Rausch, Rebecca K. Saari, and Noelle E. Selin. "A Systems Approach to Evaluating the Air Quality Co-benefits of US Carbon Policies." *Nature Climate Change* 4, no. 10 (October 2014): 917–23.

Tickner, Joel A., and Ken Geiser. "The Precautionary Principle Stimulus for Solutions- and Alternatives-Based Environmental Policy." *Environmental Impact Assessment Review* 24, no. 7–8 (October–November 2004): 801–24.

Towers, George. "Applying the Political Geography of Scale: Grassroots Strategies and Environmental Justice." *Professional Geographer* 52, no. 1 (February 2000): 23–36.

Traditional environmental organization member. Interview by the author, August 2017.

———. Interview by the author, May 2013.

TransForm and California Housing Partnership Corporation. *Why Creating and Preserving Affordable Homes near Transit Is a Highly Effective Climate Protection Strategy*. May 2014. http://www.transformca.org/sites/default/files/CHPC%20TF%20Affordable%20TOD%20Climate%20Strategy%20BOOKLET%20FORMAT.pdf.

Truong, Vien. "Addressing Poverty and Pollution: California's SB 535 Greenhouse Gas Reduction Fund." *Harvard Civil Rights-Civil Liberties Law Review* 49, no. 2 (Spring 2014): 493–529.

Tsing, Anna Lowenhaupt. *Friction: An Ethnography of Global Connection*. Princeton, NJ: Princeton University Press, 2005.

Turner, Stephen. "What Is the Problem with Experts?" *Social Studies of Science* 31, no. 1 (February 2001): 123–49.

Umemoto, Karen. "Walking in Another's Shoes: Epistemological Challenges in Participatory Planning." *Journal of Planning Education and Research* 21, no. 1 (September 2001): 17–31.

UN Framework Convention on Climate Change. "Kyoto Protocol – Targets for the First Commitment Period." Accessed March 2, 2019. https://unfccc.int/process/the-kyoto-protocol.

Union of Concerned Scientists. "Why Does CO2 Get Most of the Attention When There Are So Many Other Heat-Trapping Gases?" Last revised August 3, 2017. https://www.ucsusa.org/global-warming/science-and-impacts/science/CO2-and-global-warming-faq.html.

United Church of Christ Commission for Racial Justice. *Toxic Wastes and Race in the United States: A National Report on the Racial and Socio-economic Characteristics of Communities with Hazardous Waste Sites*. New York: Public Data Access, 1987.

Urry, John. *Climate Change and Society*. Cambridge: Polity, 2011.

US EPA. "Black Carbon Effects on Public Health and the Environment." Chap. 3 in *Report to Congress on Black Carbon*. EPA-450/R-12-001. Washington, DC: US EPA, March 2012. https://www3.epa.gov/airquality/blackcarbon/2012report/fullreport.pdf.

———. "Learn about Environmental Justice." Accessed June 11, 2018. https://www.epa.gov/environmentaljustice/learn-about-environmental-justice.

US General Accounting Office. *Siting of Hazardous Waste Landfills and Their Correlation with Racial and Economic Status of Surrounding Communities*. RCED-83-168. Washington, DC, June 14, 1983. https://www.gao.gov/products/RCED-83-168.

Van Der Heijden, Hein-Anton. "Globalization, Environmental Movements, and International Political Opportunity Structures." *Organization & Environment* 19, no. 1 (March 2006): 28–45.

Van House, Nancy. "Actor-Network Theory, Knowledge Work and Digital Libraries." Unpublished paper, 2001. http://people.ischool.berkeley.edu/~vanhouse/bridge.html.

Watts, Nick, W. Neil Adger, Paolo Agnolucci, Jason Blackstock, Peter Byass, Wenjia Cai, Sarah Chaytor, et al. "Health and Climate Change: Policy Responses to Protect Public Health." *Lancet* 386, no. 10006 (November 7, 2015): 1861–914.

Wenning, Carl J. "Scientific Epistemology: How Scientists Know What They Know." *Journal of Physics Teacher Education Online* 5, no. 2 (Autumn 2009): 3–15.

Whitmee, Sarah, Andy Haines, Chris Beyrer, Frederick Boltz, Anthony G. Capon, Braulio Ferreira de Souza Dias, Alex Ezeh, et al. "Safeguarding Human Health in the Anthropocene Epoch: Report of the Rockefeller Foundation-*Lancet* Commission on Planetary Health." *Lancet* 386, no. 10007 (November 14, 2015): 1973–2028.

———. "Testimony for the Joint Hearing by the Assembly Natural Resources Committee and the Senate Energy Committee of the California State Legislature." March 3, 2008. http://antr.assembly.ca.gov/sites/antr.assembly.ca.gov/files/hearings/Jane%20Williams%20Testimony%20-%202008.pdf.

Williams, Amanda, Steve Kennedy, Felix Philipp, and Gail Whiteman. "Systems Thinking: A Review of Sustainability Management Research." *Journal of Cleaner Production* 148 (April 1, 2017): 866–81.

Williams, Robert W. "Environmental Injustice in America and Its Politics of Scale." *Political Geography* 18, no. 1 (January 1999): 49–73.

Williamson, Rebecca. "Towards a Multi-scalar Methodology: The Challenges of Studying Social Transformation and International Migration." In *Social Transformation and Migration: National and Local Experiences in South Korea, Turkey, Mexico and Australia*, edited by Stephen Castles, Derya Ozkul, and Magdalena Arias Cubas, 17–32. London: Palgrave Macmillan, 2015.

Winkel, Jackie, Sigalle Michael, and Andrea Gordon, eds. *California's Progress toward Clean Air*. California Air Pollution Control Officers' Association, 2014. http://www.capcoa.org/wp-content/uploads/2014/04/CA_Progress_Toward_Clean_Air_Report_2014.pdf.

Wisconsin Department of Health Services. "Chemical Fact Sheet—Carbon Dioxide (CO2)." August 2013. https://www.dhs.wisconsin.gov/chemical/carbondioxide.htm.

Wold, Chris, David Hunter, and Melissa Powers. "Land Use and Forestry." Chap. 8 in *Climate Change and the Law*. 2nd ed. Newark, NJ: LexisNexis, 2013.

Wright, Laurence A., Simon Kemp, and Ian Williams. "'Carbon Footprinting': Towards a Universally Accepted Definition." *Carbon Management* 2, no. 1 (February 2011): 61–72.

Wysham, Daphne. "Carbon Market Fundamentalism." *Multinational Monitor* 29, no. 3 (November/December 2008). http://www.multinationalmonitor.org/mm2008/112008/wysham.html.

Xiang, Biao. "Multi-scalar Ethnography: An Approach for Critical Engagement with Migration and Social Change." *Ethnography* 14, no. 3 (September 2013): 282–99.

Young, Iris Marion. *Inclusion and Democracy*. Oxford: Oxford University Press, 2002.

Young, Samantha. "Schwarzenegger Opens Climate Summit with Talk by Obama." *Barre Montpelier Times Argus*, November 19, 2008. https://www.timesargus.com/articles/schwarzenegger-opens-climate-summit-with-talk-by-obama/.

Zabin, Carol, Abigail Martin, Rachel Morello-Frosch, Manuel Pastor, and James Sadd. *Advancing Equity in California Climate Policy: A New Social Contract for Low-Carbon Transition*. Center for Labor Research and Education, Donald Vial Center on Employment in the Green Economy, University of California, Berkeley, September 13, 2016. http://laborcenter.berkeley.edu/pdf/2016/Advancing-Equity.pdf.

Zaelke, Durwood, Nathan Borgford-Parnell, Stephen O. Andersen, Romina Picolotti, Dennis Clare, Xiaopu Sun, and Danielle Fest Gabrielle. "Primer on Short-Lived Climate Pollutants: Slowing the Rate of Global Warming over the Near Term by Cutting Short-Lived Climate Pollutants to Complement Carbon Dioxide Reductions for the Long Term." IGSD Working Paper. Institute for Governance and Sustainable Development, November 2013. http://www.igsd.org/documents/PrimeronShort-LivedClimatePollutantsNovemberElectronicversion.pdf.

Zimring, Carl A. *Clean and White: A History of Environmental Racism in the United States*. New York: New York University Press, 2015.

Zuckerman, Adam, and Kevin Koenig. *From Well to Wheel: The Social, Environmental, and Climate Costs of Amazon Crude*. Oakland, CA: Amazon Watch, September 2016. https://amazonwatch.org/assets/files/2016-amazon-crude-report.pdf.

Index

Figures, notes, and tables are indicated by *f*, n, and *t* following the page number.